SETS FOR MATHEMATICS

Advanced undergraduate or beginning graduate students need a unified foundation for their study of mathematics. For the first time in a text, this book uses categorical algebra to build such a foundation, starting from intuitive descriptions of mathematically and physically common phenomena and advancing to a precise specification of the nature of categories of sets.

Set theory as the algebra of mappings is introduced and developed as a unifying basis for advanced mathematical subjects such as algebra, geometry, analysis, and combinatorics. The formal study evolves from general axioms that express universal properties of sums, products, mapping sets, and natural number recursion. The distinctive features of Cantorian abstract sets, as contrasted with the variable and cohesive sets of geometry and analysis, are made explicit and taken as special axioms. Functor categories are introduced to model the variable sets used in geometry and to illustrate the failure of the axiom of choice. An appendix provides an explicit introduction to necessary concepts from logic, and an extensive glossary provides a window to the mathematical landscape.

SETS FOR MATHEMATICS

F. WILLIAM LAWVERE
State University of New York at Buffalo

ROBERT ROSEBRUGH
Mount Allison University

CAMBRIDGE
UNIVERSITY PRESS

CAMBRIDGE
UNIVERSITY PRESS

32 Avenue of the Americas, New York NY 10013-2473, USA

Cambridge University Press is part of the University of Cambridge.

It furthers the University's mission by disseminating knowledge in the pursuit of education, learning and research at the highest international levels of excellence.

www.cambridge.org
Information on this title: www.cambridge.org/9780521010603

First published 2003

A catalogue record for this publication is available from the British Library

Library of Congress Cataloguing in Publication data
Lawvere, F. W.
Sets for mathematics / F. William Lawvere, Robert Rosebrugh.
p.   cm.
Includes bibliographical references and index.
ISBN 0-521-80444-2 – ISBN 0-521-01060-8 (pbk.)
1. Set theory.   I. Rosebrugh, Robert, 1948–   II. Title.
QA248 .L28   2002
511.3′22 – dc21   2002071478

ISBN 978-0-521-80444-8 Hardback
ISBN 978-0-521-01060-3 Paperback

Portraits on the front cover are of Georg Cantor and Richard Dedekind (top) and Samuel Eilenberg and Saunders Mac Lane (bottom). The portrait of Samuel Eilenberg appears by kind permission of Columbia University.

Contents

Foreword

Why *Sets for Mathematics*?

This book is for students who are beginning the study of advanced mathematical subjects such as algebra, geometry, analysis, or combinatorics. A useful foundation for these subjects will be achieved by openly bringing out and studying what they have in common.

A significant part of what is common to all these subjects was made explicit 100 years ago by Richard Dedekind and Georg Cantor, and another significant part 50 years ago by Samuel Eilenberg and Saunders Mac Lane. The resulting idea of **categories of sets** is the main content of this book. It is worth the effort to study this idea because it provides a unified guide to approaching constructions and problems in the science of space and quantity.

More specifically, it has become standard practice to represent an object of mathematical interest (for example a surface in three-dimensional space) as a "structure." This representation is possible by means of the following two steps:

(1) First we deplete the object of nearly all content. We could think of an idealized computer memory bank that has been erased, leaving only the pure locations (that could be filled with any new data that are relevant). The bag of pure points resulting from this process was called by Cantor a *Kardinalzahl*, but we will usually refer to it as an **abstract set**.

(2) Then, just as computers can be wired up in specific ways, suitable specific **mappings** between these structureless sets will constitute a structure that reflects the complicated content of a mathematical object. For example, the midpoint operation in Euclidean geometry is represented as a mapping whose "value" at any pair of points is a special third point.

To explain the basis for these steps there is an important procedure known as the **axiomatic method**: That is, from the ongoing investigation of the ideas of sets and mappings, one can extract a few statements called *axioms*; experience has shown that these axioms are sufficient for deriving most other true statements by pure logic when that is useful. The use of this axiomatic method makes naive set theory rigorous and helps students to master the ideas without superstition. An analogous procedure was applied by Eilenberg and Steenrod to the ongoing development of algebraic topology in their 1952 book [ES52] on the foundations of that subject as well as by other practitioners of mathematics at appropriate stages in its history.

Some of the foundational questions touched on here are treated in more detail in the glossary (Appendix C) under headings such as Foundations, Set Theory, Topos, or Algebraic Topology.

Organization

In Chapters 1–5 the emphasis is on the category of abstract sets and on some very simple categorical generalities. The additional century of experience since Cantor has shown the importance of emphasizing some issues such as:

(1) Each map needs both an explicit domain and an explicit codomain (not just a domain, as in previous formulations of set theory, and not just a codomain, as in type theory).

(2) Subsets are not mere sets with a special property but are explicit inclusion maps. (This helps one to realize that many constructions involving subsets are simplified and usefully generalized when applied appropriately to maps that are not necessarily subsets.)

(3) The algebra of composition satisfies the familiar associative and identity rules; other basic concepts, such as "belonging to" (e.g., membership in, and inclusion among, subsets) and the dual "determined by" are easily expressible as "division" relative to it. It turns out that this adherence to algebra (showing that "foundation" does not need a language distinct from that of ordinary mathematics) has broad applicability; it is particularly appropriate in smoothing the transition between constant and variable sets.

(4) Because functionals play such a key role in mathematics, the algebra is explicitly strengthened to include the algebra of evaluation maps and induced maps.

All of these issues are elementary and quite relevant to the learning of basic mathematics; we hope that mathematics teachers, striving to improve mathematics education, will take them to heart and consider carefully the particular positive role that explicit formulations of mathematical concepts can play.

Beginning in Chapter 6, examples of categories of cohesive and variable sets are gradually introduced; some of these help to objectify features of the constant sets such as recursion and coequalizers.

We illustrate the use of the maximal principle of Max Zorn in Appendix B, and we include a proof of it in the form of exercises with hints. Several other results that do not hold in most categories of variable or cohesive sets, such as the Schroeder–Bernstein theorem and the total ordering of sizes, are treated in the same way. Despite our axiomatic approach, we do not use the internal language that some books on topos theory elaborate; it seemed excessively abstract and complicated for our needs here.

Appendix A presents essentially a short course in "all that a student needs to know about logic." Appendix B, as mentioned, briefly treats a few of the special topics that a more advanced course would consider in detail. Appendix C provides a glossary of definitions for reference. Some of the glossary entries go beyond bare definition, attempting to provide a window into the background.

Some exercises are an essential part of the development and are placed in the text, whereas others that are optional but recommended are for further clarification.

F. William Lawvere   Robert Rosebrugh

June 2002

Contributors to Sets for Mathematics

This book began as the transcript of a 1985 course at SUNY Buffalo and still retains traces of that verbal record. Further courses in the 1990s at Buffalo and at Mount Allison followed. We are grateful to the students in all those courses for their interest, patience, and careful reading of the text, and in particular to the late Ed Barry for his detailed set of notes and his bibliographical material.

John Myhill made the original course possible and also contributed some incisive observations during the course itself. Max Zorn sent an encouraging postcard at a critical juncture. Saunders Mac Lane strongly urged that the original transcript be transformed into a book; we hope that, sixteen years later, the result of the transformation will approach fulfillment of his expectations.

Several people contributed to making this work a reality, starting with Xiao-Qing Meng, who was the original teaching assistant. By far the most creative and inspiring contributor has been Fatima Fenaroli, who worked tirelessly behind the scenes from the very beginning, and without whose constant collaboration the project would never have come so far. Indispensable have been the advice and support of Steve Schanuel; his mathematical insight, pedagogical sense, and unfailing ability to spot errors have assisted through the many stages of revision.

Ellen Wilson typed the first TEX version, which was the basis for further revisions. Some of those revisions were made in response to constructive criticism by Giuseppe Rosolini and Richard Wood, who gave courses (at Genoa and Dalhousie, respectively) using the draft manuscript, and by the anonymous referees. Federico Lastaria and Colin McLarty studied the manuscript and contributed several improvements. The drawings in xy-pic were created by Francisco Marmolejo. The transformation into the current volume was made possible by our editor Roger Astley and by Eleanor Umali, who managed the production process.

We are grateful to Rebecca Burke for her patient, understanding encouragement that was crucial to the completion of this work, and for her hospitality, which made our collaboration enjoyable.

All of these people have contributed toward the goal of concentrating the experience of the twentieth century in order to provide a foundation for twenty-first century education and research. Although our effort is only one of the first steps in that program, we sincerely hope that this work can serve as a springboard for those who carry it further.

F. William Lawvere and Robert Rosebrugh

1

Abstract Sets and Mappings

1.1 Sets, Mappings, and Composition

Let us discuss the idea of abstract constant sets and the mappings between them in order to have a picture of this, our central example, before formalizing a mathematical definition. An abstract set is supposed to have elements, each of which has no structure, and is itself supposed to have no internal structure, except that the elements can be distinguished as equal or unequal, and to have no external structure except for the number of elements. In the category of abstract sets, there occur sets of all possible sizes, including finite and infinite sizes (to be defined later). It has been said that an abstract set is like a mental "bag of dots," except of course that the bag has no shape; thus,

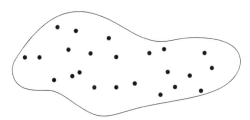

may be a convenient way of picturing a certain abstract set for some considerations, but what is apparently the same abstract set may be pictured as

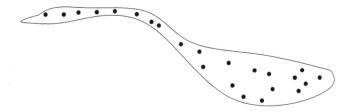

for other considerations.

What gives the category of sets its power is the concept of **mapping**. A mapping f from an abstract set A to an abstract set B is often explained through the use of the word *value*. (However, since the elements of B have no structure, it would be misleading to always think of these values as quantities.) Each mapping f from A to B satisfies

for each element x of A

there is exactly one element y of B

such that y is a value of f at x

This justifies the phrase "*the* value"; the value of f at x is usually denoted by $f(x)$; it is an element of B. Thus, a mapping is single-valued and everywhere defined (everywhere on its domain) as in analysis, but it also has a *definite codomain* (usually bigger than its set of actual values). *Any* f at all that satisfies this one property is considered to be a mapping from A to B in the category of abstract constant sets; that is why these mappings are referred to as "arbitrary". An important and suggestive notation is the following:

Notation 1.1: *The arrow notation* $A \xrightarrow{f} B$ *just means the* **domain** *of* f *is* A *and the* **codomain** *of* f *is* B, *and we write* $dom(f) = A$ *and* $cod(f) = B$. (We will usually use capital letters for sets and lowercase letters for mappings.) For printing convenience, in simple cases this is also written with a colon $f : A \longrightarrow B$. We can regard the notation $f : A \longrightarrow B$ as expressing the statement $dom(f) = A$ & $cod(f) = B$, where & is the logical symbol for *and*.

For small A and B, a mapping from A to B can be pictured using its cograph or internal diagram by

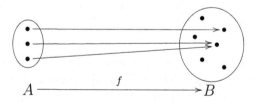

where $f(x)$ is the dot at the right end of the line that has x at its left end for each of the three possible elements x.

Abstract sets and mappings are a **category**, which means above all that there is a **composition** of mappings, i.e., given any pair $f : A \longrightarrow B$ and $g : B \longrightarrow C$ there is a specified way of combining them to give a resulting mapping $g \circ f : A \longrightarrow C$. Note that the codomain set of the first mapping f must be *exactly the same set* as the domain set of the second mapping g. It is common to use the notation $\circ$ for composition and to read it as "following," but we will also, and much more often, denote the composite "*g* following *f*" just by *gf*. A particular instance of composition can be pictured by an external diagram or by an internal diagram as below. First consider any three mappings *f*, *g*, and *m* with domains and codomains as indicated:

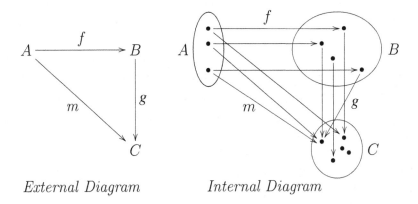

External Diagram   *Internal Diagram*

The internal cograph diagrams express the full information about particular maps, which is often more than we need; thus, we will use simple, external diagrams wherever possible.

Since any mapping satisfies restrictions of the kind "for each . . . there is exactly one . . . ," in the diagram above, we observe that

- for each element *a* of *A* there is exactly one element *b* of *B* for which *b* is a value of *f* at *a* (briefly *f*(*a*) = *b*);
- for each element *b* of *B* there is exactly one element *c* of *C* for which *c* is a value of *g* at *b* (briefly *g*(*b*) = *c*);
- for each element *a* of *A* there is exactly one element *c* of *C* for which *c* is a value of *m* at *a* (briefly *m*(*a*) = *c*).

The external diagram above is said to be a "commutative diagram", if and only if *m* is actually the composite of *g* *following* *f*; then, notationally, we write simply *m* = *gf*.

More precisely, for the triangular diagram to be considered commutative, the relation between *f*, *g*, *m* must have the following property:

For each element *a* of *A* we can find the value of *m*(*a*) by proceeding in two steps: first find *f*(*a*) and then find *g*(*f*(*a*)); the latter is the *same* as *m*(*a*).

(Examining the internal diagram shows that *m* = *gf* in the figure above.)

A familiar example, when *A* = *B* = *C* is a set of numbers equipped with structural mappings providing addition and multiplication, involves $f(x) = x^2$ and $g(x) = x + 2$ so that $(g \circ f)(x) = x^2 + 2$. The value of the composite mapping at *x* is the result of taking the value of *g* at the value of *f* at *x*. In contexts such as this where both multiplication and composition are present, it is necessary to use distinct notations for them.

Exercise 1.2

Express the mapping that associates to a number x the value $\sqrt{x^2 + 2}$ as a composite of *three* mappings.   ◊

We need to be more precise about the concept of category. The ideas of set, mapping, and composition will guide our definition, but we need one more ingredient. For each set A there is the **identity mapping** $1_A : A \longrightarrow A$ whose values are determined by $1_A(x) = x$. For any set A, this definition determines a particular mapping among the (possibly many) mappings whose domain and codomain are both A.

On the basis of the preceding considerations we have part of the information required to define the general notion of "category". The first two items listed correspond to abstract sets and arbitrary mappings in the example of the category of sets.

A category C has the following data:

- Objects: denoted $A, B, C, \ldots$
- Arrows: denoted $f, g, h, \ldots$ (arrows are also often called *morphisms* or *maps*)
- To each arrow f is assigned an object called its *domain* and an object called its *codomain* (if f has domain A and codomain B, this is denoted $f : A \longrightarrow B$)
- Composition: To each $f : A \longrightarrow B$ and $g : B \longrightarrow C$ there is assigned an arrow $gf : A \longrightarrow C$ called *"the composite of f and g"* (or *"g following f"*)
- Identities: To each object A is assigned an arrow $1_A : A \longrightarrow A$ called *"the identity on A"*.

1.2 Listings, Properties, and Elements

We have not finished defining *category* because the preceding data must be constrained by some general requirements. We first continue with the discussion of elements. Indeed, we can immediately simplify things a little: an idea of element is not necessary as a *separate* idea because we may always identify the elements themselves as special mappings. That will be an extreme case of the *parameterizing* of elements of sets. Let us start with a more intermediate case, for example, the set of mathematicians, *together with* the indication of two examples, say Sir Isaac Newton and Gottfried Wilhelm Leibniz. Mathematically, the model will consist not only of an abstract set A, (to stand for the set of all mathematicians) but also of another abstract set of two elements 1 and 2 to act as labels *and* the specified mapping with codomain A whose value at 1 is "Newton" and whose value at 2 is "Leibniz". The two-element set is the domain of the parameterization.

Such a specific *parameterization* of elements is one of two kinds of features of a set ignored or held in abeyance when we form the *abstract* set. Essentially, all of the terms – **parameterization, listing, family** – have abstractly the same meaning: simply looking at one *mapping* into a set A of interest, rather than just at the one set A all by itself.

Whenever we need to insist upon the abstractness of the sets, such a preferred listing is one of the two kinds of features we are abstracting away.

The other of the two aspects of the elements of an actual concrete aggregation (which are to be ignored upon abstraction) involves the **properties** that the elements might have. For example, consider the set of all the mathematicians and the property "was born during the seventeenth century" that some of the mathematicians have and some do not. One might think that this is an important property of mathematicians as such, but nonetheless one might momentarily just be interested in how many mathematicians there are.

Certain properties are interpreted as particular mappings by using the two-element set of "truth values" – true, false – from which we *also* arrive (by the abstraction) at the abstract set of two elements within which "true" could be taken as exemplary. If we consider a particular mapping such as

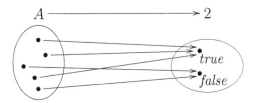

we see that all those elements of A that go to "true" will constitute one portion of A, and so f determines a property "true" for some elements, and "not true," or "false," for others. There are properties for which the codomain of f will need more than two elements, for example, age of people: the codomain will need at least as many elements as there are different ages.

As far as listing or parameterizing is concerned, an extreme case is to imagine that *all* the elements have been listed by the given procedure. The opposite extreme case is one in which *no* examples of elements are being offered even though the actual set A under discussion has some arbitrary size. That is, in this extreme case the index set is an *empty* set. Of course, the whole listing or parameterization in this extreme case amounts really to nothing more than the one abstract set A itself.

Just short of the extreme of not listing any is listing just one element. We can do this using a one-element set as parameter set.

To characterize mathematically what the one-element set is, we will consider it in terms of the property that does not distinguish. The following is the first axiom we require of the category of sets and mappings.

AXIOM: TERMINAL SET
There is a set 1 such that for any set A there is exactly one mapping $A \longrightarrow 1$. This unique mapping is given the same name A as the set that is its domain.

We call 1 a **terminal object** of the category of sets and mappings. There may or may not be more than one terminal object; it will make no difference to the mathematical content. In a given discussion the symbol 1 will denote a chosen terminal object; as we will see, which terminal object is chosen will also have no effect on the mathematical content.

Several axioms will be stated as we proceed. The axiom just stated is part of the stronger requirement that the category of sets and mappings has finite inverse limits (see Section 3.6). A typical cograph picture is

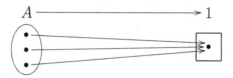

Only a one-element set $V = 1$ can have the extreme feature that one cannot detect any distinctions between the elements of A by using only "properties" $A \longrightarrow V$. Having understood what a one-element set is in terms of mapping *to* it, we can now use mappings *from* it to get more information about arbitrary A.

Definition 1.3: *An* **element** *of a set A is any mapping whose codomain is A and whose domain is 1 (or abbreviated ... $1 \xrightarrow{a} A$).*

(Why does 1 itself have exactly one element according to this definition?)
The first consequence of our definition is that

element **is a special case of mapping.**

A second expression of the role of 1 is that

evaluation **is a special case of composition.**

In other words, if we consider any mapping f from A to B and then consider any element a of A, the codomain of a and the domain of f are the same; thus, we can

form the composite fa,

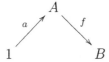

which will be a mapping $1 \longrightarrow B$. But since the domain is 1, this means that fa is an *element* of B. Which element is it? It can only be, and clearly is, the *value* of f at a:

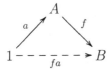

That is, if a is an element, $fa = f(a)$.

Finally, a third important expression of the role of 1 is that

evaluation of a composite is a special case of the
Associative law

of composition (which will be one of the clauses in the definition of *category*). In order to see this, suppose $m = gf$ and consider

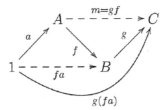

The formula (in which we introduce the symbols $\forall$ to mean "for all" and $\Rightarrow$ to mean "implies")

$$m = gf \implies [\forall a [1 \xrightarrow{a} A \Rightarrow m(a) = g(fa)]]$$

expresses our idea of evaluation of the composition of two mappings; i. e. if m is the composite of f and g, then for any element a of the domain of f the value of m at a is equal to the value of g at $f(a)$. More briefly, $(gf)a = g(fa)$, which is a case of the associative law.

The three points emphasized here mean that our internal pictures can be (when necessary or useful) completely interpreted in terms of external pictures by also using the set 1.

Notice that the axiom of the terminal set and the definition of element imply immediately that the set 1 whose existence is guaranteed by the axiom has *exactly* one element, namely, the unique mapping from 1 to 1. There is always an identity mapping from a set to itself, so this unique mapping from 1 to 1 must be the identity mapping on 1.

We want to introduce two more logical symbols: the symbol $\exists$ is read "there exists," and $\exists!$ is read "there exists exactly one". Thus, we can repeat the characteristic feature of every $f : A \longrightarrow B$ as follows:

$$\forall a : 1 \longrightarrow A \ \exists! \, b : 1 \longrightarrow B[b \text{ is a value of } f \text{ at } a]$$

But this is a special case of the fact that composition in general is uniquely defined.

1.3 Surjective and Injective Mappings

Recall the first internal diagram (cograph) of a mapping that we considered:

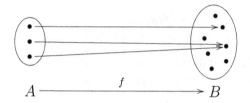

Note that it is *not* the case for the f in our picture that

for each element b of B
there is an element x of A
for which b is the value of f at x. ($f(x) = b$)

Definition 1.4: *A mapping $f : A \longrightarrow B$ that has the existence property "for each element b of B there is an element x of A for which $b = f(x)$" is called a* **surjective mapping**.

Neither is it the case that the f in our picture has the property

for each element b of B
there is at most one element x of A
for which $f(x) = b$

Definition 1.5: *A mapping $f : A \longrightarrow B$ that has the uniqueness property "given any element b of B there is at most one element x of A for which $f(x) = b$" is called an* **injective mapping**. *In other words, if f is an injective mapping, then for all elements x, x' of A, if $f(x) = f(x')$, then $x = x'$.*

Definition 1.6: *A mapping that is both surjective and injective is called* **bijective**.

Thus, the f pictured above is neither surjective nor injective, but in the figure below $g : A \longrightarrow B$ is an *injective* mapping from the same A and to the same B.

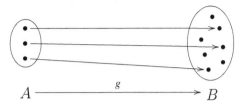

Exercise 1.7
Is the pictured g surjective? ◊

Exercise 1.8
Are there any surjective mappings $A \longrightarrow B$ for the pictured A, B? ◊

Exercise 1.9
How many mappings from the three-element set A to the seven-element set B are there? Can we picture them all? ◊

Exercise 1.10
Same as 1.9, but for mappings $B \longrightarrow A$ from a seven-element to a three-element set. ◊

Exercise 1.11
Are there any surjective $B \longrightarrow A$? Are there any injective ones? ◊

Exercise 1.12
What definition of "$f_1 \neq f_2$" is presupposed by the idea "number of" mappings we used in 1.9 and 1.10? ◊

Exercises 1.9 and 1.12 illustrate that the feature "external number/internal inequality of instances" characteristic of an abstract set is also associated with the notion "mapping from A to B," except that the elements (the mappings) are not free of structure. But abstractness of the sets really means that the elements are for the moment considered without internal structure. By considering the mappings from A to B with their internal structure ignored, we obtain a new abstract set B^A. Conversely, we will see in Chapter 5 how any abstract set F of the right size can act as mappings between given abstract sets. (For example, in computers variable programs are just a particular kind of variable data.)

1.4 Associativity and Categories

Recall that we saw in Section 1.2 that an "associative law" in a special case expresses the evaluation of composition. Indeed, whenever we have

$$1 \xrightarrow{\ a\ } A \xrightarrow{\ f\ } B \xrightarrow{\ g\ } C$$

then we have the equation $(gf)(a) = g(fa)$.

If we replace a by a general mapping $u : T \longrightarrow A$ whose domain is not necessarily 1, we obtain the **Associative law**

$$(gf)u = g(fu)$$

which actually turns out to be true for any three mappings that can be composed; i.e., that from the commutativity of the two triangles below we can conclude that moreover the outer two composite paths from T to C have equal composites (it is said that the whole diagram is therefore "commutative").

$$(gf)u = g(fu)$$
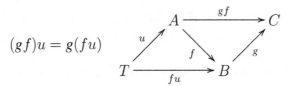

Since the 1 among abstract sets has the special feature (which we discuss in Section 1.5) that it can *separate* mappings, in abstract sets the general associative law follows from the special case in which $T = 1$.

An important property of identity mappings is that they not only "do nothing" to an element but that they have this same property with respect to composition. Thus, if $1_A : A \longrightarrow A$ and $1_B : B \longrightarrow B$ are identity mappings, then for any $f : A \longrightarrow B$ we have the equations

$$f1_A = f = 1_B f$$

With these ideas in hand we are ready to give the completed definition of category. The beginning of our specification repeats what we had before:

Definition 1.13: *A* **category** $\mathcal{C}$ has the following data:

- Objects: denoted $A, B, C, \ldots$
- Arrows: denoted $f, g, h, \ldots$ (arrows are also often called **morphisms** or **maps**)
- To each arrow f is assigned an object called its **domain** and an object called its **codomain** (if f has domain A and codomain B, this is denoted $f : A \longrightarrow B$ or $A \xrightarrow{\ f\ } B$)
- Composition: To each $f : A \longrightarrow B$ and $g : B \longrightarrow C$, there is assigned an arrow $gf : A \longrightarrow C$ called **"the composite g following f"**
- Identities: To each object A is assigned an arrow $1_A : A \longrightarrow A$ called **"the identity on A"**.

The data above satisfy the axioms

- Associativity: if $A \xrightarrow{f} B \xrightarrow{g} C \xrightarrow{h} D$, then $h(gf) = (hg)f$
- Identity: if $f : A \longrightarrow B$, then $f = f1_A$ and $f = 1_B f$.

As we have been emphasizing,

AXIOM: S IS A CATEGORY
Abstract sets and mappings form a category (whose objects are called sets, and whose arrows are called mappings).

This is the basic axiom implicit in our references to the "category of abstract sets and mappings" above. There are many other examples of categories to be found in mathematics, and a few of these are described in the exercises in Section 1.8 at the end of the chapter.

1.5 Separators and the Empty Set

If a pair of mappings

$$A \underset{f_2}{\overset{f_1}{\rightrightarrows}} B$$

has the same domain and has the same codomain (i.e., they are two mappings that *could* be equal), then we can discover whether they are really equal by testing with elements

$$(\forall x[1 \xrightarrow{x} A \Rightarrow f_1 x = f_2 x]) \Longrightarrow f_1 = f_2$$

i.e., if the value of f_1 equals the value of f_2 at every element x of A, then $f_1 = f_2$. This is one of the ways in which we can conclude that $f_1 = f_2$. The converse implication of the statement is trivial because it is merely substitution of equals for equals (a general idea of mathematics). But the indicated implication is a special, particularly powerful feature of one-element abstract sets. In its *contrapositive* form it states: If $f_1 \neq f_2$, then there exists at least one element x at which the values of f_1 and f_2 are different. (This is the answer to Exercise 1.12!) For a category C an object with this property is called a *separator.*

Definition 1.14: *An object S in a category C is a **separator** if and only if whenever*

$$X \underset{f_2}{\overset{f_1}{\rightrightarrows}} Y$$

are arrows of C then

$$(\forall x [S \xrightarrow{x} X \Rightarrow f_1 x = f_2 x]) \Longrightarrow f_1 = f_2$$

As mentioned in 1.4 the property we have been describing is *required* of the terminal object 1 as a further axiom in the category of abstract sets and arbitrary mappings. It is a powerful axiom with many uses; it is special to the category S of abstract sets and will not hold in categories of variable and cohesive sets where more general elements than just the "points" considered here may be required for the validity of statements even analogous to the following one (see Section 1.6):

AXIOM: THE TERMINAL OBJECT 1 SEPARATES MAPPINGS IN S
A one-element set 1 *is a* separator *in S, i.e., if*

$$X \overset{f_1}{\underset{f_2}{\rightrightarrows}} Y$$

then

$$(\forall x\, [1 \xrightarrow{\ x\ } X \Rightarrow f_1 x = f_2 x]) \Longrightarrow f_1 = f_2$$

Exercise 1.15
In the category of abstract sets S, any set A with at least one element $1 \xrightarrow{\ x\ } A$ is also a separator. (When an exercise is a statement, prove the statement.)   ◊

We return to the extreme case of listing or parameterization in which no elements are listed. In this case there cannot be more than one listing map (we will use "map" and "mapping" synonymously!) into A since the indexing set we are trying to use is empty. On the other hand, there must be one since the statement defining the property of a mapping is a requirement on each element of the domain set (that there is assigned to it a unique value element in the codomain). This property is satisfied "vacuously" by a mapping from a set without elements since there is simply no requirement. Thus, there exists a unique mapping from an empty set to any given set. We require such a set as an axiom.

AXIOM: INITIAL SET
There is a set 0 *such that for any set A there is exactly one mapping* $0 \longrightarrow A$.

We call 0 an **initial object** of the category of sets and mappings.

Note that the form of this axiom is the same as the form of the axiom of the terminal set, i.e. we require the existence of a set and a unique mapping for every set *except* that the unique mapping is now *to* the arbitrary set whereas formerly it was *from* the arbitrary set. Like the axiom of the terminal set, the axiom of the initial set will become part of a stronger axiom later. The initial set is often called the *empty set* because, as we will later see, there are no maps $1 \longrightarrow 0$.

Exercise 1.16

In the category of abstract sets S the initial set 0 is not a separator. (Assume that two sets A and B exist with at least two maps $A \longrightarrow B$.) $\Diamond$

ADDITIONAL EXAMPLES:

(1) If T is an index set of numbers, then

$$T \xrightarrow{\ x\ } X$$

could be the listing of all the American presidents in chronological order. It does turn out that the map is not injective – Cleveland was both the 22nd and the 24th president.

If we want to ask who was the 16th president, the structure of the question involves all three: the actual set, the actual listing, and a choice of index:

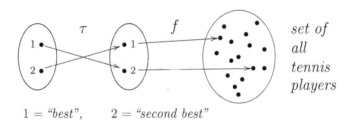

Lincoln derives by composing the index $i = 16$ and the list x of presidents.

(2) There are at least two uses of two-element sets:

Index sets and *truth-value sets*

Consider

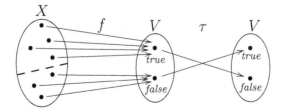

$1 = $ "*best*", $2 = $ "*second best*"

The one that used to be the second-best tennis player could become the best; encode that by noting that there is an **endomapping** (or self-mapping) τ that interchanges the two denominations. The list f' that is correct today can be the reverse of the list f that was true yesterday if an "upset" match occurred; i.e. we could have $f' = f\tau$.

A similar sort of thing happens also on the side of the possible properties of the elements of X:

which results in

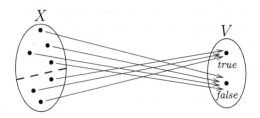

In this case we could also compose with the τ, but now it would instead be τ following f (which is written τf). This is called logical negation since it transforms f into *not-f*, i.e. *(not-f)(x) = not-f(x)*. The composite property is the property of not having the property f. Often in the same discussion both reparameterization of lists and logical or arithmetic operations on properties occur, as suggested in the following diagram:

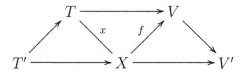

If we have a list x of elements and a property f, then the composite fx can be thought of in two equally good ways. Because V represents values, we can think of this fx as just a property of elements of T; for example, given the listing x of the presidents, the property f of their being Democrats becomes a property of indices. But fx could also be considered as a list (of truth values). The two concepts thus reduce to the same in the special case $T \longrightarrow V$, giving

LIST		TRUTH VALUES		PROPERTY		INDICES
or	of	or	=	or	of	or
FAMILY		QUANTITIES		MEASUREMENT		PARAMETERS

Of course, the words for T (indices/parameters) and the words for V (truthvalues/quantities) only refer to structure, which is "forgotten" when T, V are abstract sets (but which we will soon "put back in" in a more conscious way); we mention this fact mainly to emphasize its usefulness (via specific x and f) even when the structure forgotten on X itself was of *neither* of those kinds.

Exercise 1.17
Consider

$$S = \text{Set of socks in a drawer in a dark room}$$
$$V = \{\text{white, black}\}$$
$$f = \text{color}$$

How big must my "sampler" T be in order that for *all injective* x, fx is *not injective* (i.e., at least two chosen socks will be "verified" to have the *same* color)?

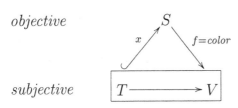

objective

subjective

◊

1.6 Generalized Elements

Consider the following three related statements (from Sections 1.2 and 1.4):

(1) *Element* is a special case of mapping;
(2) *Evaluation* is a special case of composition;
(3) *Evaluation of a composite* is a special case of the associative law of composition.

Statement (2) in one picture is

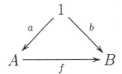

(that is to say, $fa = b$) in which a, b are elements considered as a special case of the commutativity of the following in which a, b are general mappings:

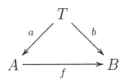

"Taking the value" is the special case of composition in which T is taken to be 1.

For statement (3), recall that the associative law applies to a situation in which we have in general three mappings:

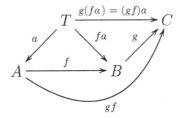

We can compute the triple composite in two ways: We can either form fa and follow that by g, getting $g(fa)$, or we can first form gf (g following f) and consider a followed by that, obtaining what we call $(gf)a$; the associative law of composition says that these are always equal for any three mappings a, f, g.

In a special case, where $T = 1$, the description would be that the value of the composite gf at an element a is by definition the element of C that we get by successive evaluation of f and then of g, leading to the same picture

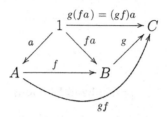

but one that is special because it emanates from the one-element set.

In many cases we will actually want to reverse this specialization procedure; that is, by the phrase

an element of A

we often actually mean a mapping to A from any T (not only $T = 1$). In case of confusion we may refer to this as

a generalized element of A

A generalized element of A is actually nothing but an arbitrary mapping whose codomain happens to be A. It has some domain, but we do not insist that that domain be 1.

A more accurate description in terms of the content would be

variable element of A varying over T

This is a very ancient way of using the idea of element; for example, if we consider the way people talk about temperature, they say *the* temperature, which seems to be an element of the set of temperatures, and intend by it the actual temperature. On the other hand, yesterday it was one value, and today it is another value. It is varying, but it is still *the*. Just think of

T as the set of days

for every day there is an element in the constant sense. The actual temperature on that day is the value of a mapping; the mapping itself is the temperature the weatherman is talking about. It is varying, but it is still considered as one entity, one "element". We speak this way about many other situations. "I am a different person today than I was yesterday, yet I am still the same person." To explain anything that involves that sort of variation, one very common kind of model for it will involve somehow (not as its only ingredient by any means but as one ingredient)

- an abstract set that plays the role of the "temporal" parameters,
- another one that plays the role of values, and
- a specific mapping that describes the result of the evolution process.

The process itself has some cause that may also have to be mentioned, but the result of that cause will be a succession of values; change is often described that way. But we will always correctly persist in attaching the definite article "the" as in "the temperature," "the me" to express the underlying unity. An element of the set of temperatures might very well be

(a) one single unchanging value of temperatures;

equally well it might be

(b) the temperature of Milan during one week,

in which case it is still one (generalized) element of the set of temperatures, but its domain is T.

Variable *sets* also occur in real life; for example, the set of people inhabiting a particular house may vary from year to year. To mathematically model such situations, categories of variable sets are useful. In this chapter we are emphasizing the category in which the sets themselves are constant, but later we will explicitly describe and construct examples of categories in which the sets are (even continuously) variable.

1.7 Mappings as Properties

What about the mappings that have domain A? They certainly are not elements of A in the usual sense; they could be called *properties* of (elements of) A.

To deal first with some trivial cases, let us review the definition of 1 and 0. The characterizing features of an empty set 0 is that

$$\text{for each set } A \text{ there is just one mapping } 0 \longrightarrow A,$$

whereas the characterizing features of a terminal set 1 is

$$\text{for each set } A \text{ there is just one mapping } A \longrightarrow 1.$$

The descriptions of 0 and 1 look rather similar except that the arrows are reversed, but we will see that these objects are quite different when we start talking about mappings from 1 and into 0 such as

$$1 \longrightarrow B \quad \text{versus} \quad A \longrightarrow 0$$

Whereas mappings from $1 \longrightarrow B$ exist for any B that is nonempty – and there will be many of them (depending on the size of B) – by contrast $A \longrightarrow 0$ will exist only if A is also empty, but even then there is only one map.

So the sets whose existence is required by the axioms of the terminal set and the initial set should be very different! To allow $0 = 1$ would result in *all* sets having only one (narrow sense) element. We must clearly avoid that. For sets A and B, the notation $A \cong B$, which is read "A is isomorphic to B," means that there are mappings $f : A \longrightarrow B$ and $g : B \longrightarrow A$ satisfying $gf = 1_A$ and $fg = 1_B$. We will have much more to say about this concept later (see Section 3.2), but for now we use it (in a negative way) in the following:

AXIOM: NONDEGENERACY OF $\mathcal{S}$

$$0 \ncong 1$$

Exercise 1.18
How many mappings are there from 0 to 1? from 0 to 0? from 1 to 0? (and so how many elements does 0 have?)

Hint: To answer the third question requires more axioms than do the other two.

$\Diamond$

Notice that we could have taken the apparently stronger statement "there is no mapping from 1 to 0" instead of the axiom as stated. The stronger statement certainly implies the axiom, but as shown by the exercise the axiom implies the stronger statement too.

We are considering the question, If V is a fixed set of values, then for any A, what kind of properties $A \longrightarrow V$ can there be? ("Property" is being used in such a general sense that it means just an arbitrary mapping but from the dual point of view to that of generalized elements.)

If V is 0, we see that the answer to the preceding question is "none," unless A is itself empty, and in that case only one. If $V = 1$, there is exactly one mapping for any A. We have to take a V that is somewhat larger if we want any interesting property of (elements of) A at all. Thus, we will take a *two*-element set for V and see that we in fact get enough properties to "discriminate" between the elements of any A.

The ability of "discriminating" elements by means of V-valued properties is frequently expressed by the following:

Definition 1.19: *An object V is a* **coseparator** *if for any A and for any parallel pair*

$$T \underset{a_1}{\overset{a_0}{\rightrightarrows}} A$$

(of "generalized elements")

$$(\forall \varphi [A \xrightarrow{\varphi} V \implies \varphi a_0 = \varphi a_1]) \implies [a_0 = a_1]$$

Notice that if V is a coseparator, then $a_0 \neq a_1$ entails that there is $A \xrightarrow{\varphi} V$ with $\varphi a_0 \neq \varphi a_1$, i.e. V can discriminate elements. By what we have said here, neither $V = 0$ nor $V = 1$ can coseparate.

Exercise 1.20
Use the fact that 1 is a separator in the category of abstract sets to show that (in that category), if V can discriminate elements, then it can discriminate generalized elements. Thus, in $\mathcal{S}$ the general T in Definition 1.19 could safely be replaced by $T = 1$. ◇

Claim 1.21: *If V is a set with exactly two elements, call it 2, then V is a coseparator in the category of abstract sets.*

Remark 1.22: We are assuming that there exists a set 2 with exactly two elements: $1 \underset{1}{\overset{0}{\rightrightarrows}} 2$. We will make this existence more precise in Chapter 2. We also denote by $\tau : 2 \longrightarrow 2$ the mapping completely defined by $\tau(0) = 1$ and $\tau(1) = 0$.

Here is a provisional justification of Claim 1.21 based on our naive conception of $V = 2$; soon we will have enough axioms to formally prove it. Assume that A is *arbitrary* but that there are maps $1 \underset{a_1}{\overset{a_0}{\rightrightarrows}} A$ with $a_0 \neq a_1$. We must "construct" a $A \xrightarrow{\varphi} 2$ with $\varphi a_0 \neq \varphi a_1$ showing that 2 coseparates (which follows by using the result of Exercise 1.20). That this is possible is quite obvious from a picture, as follows. If A is an arbitrary abstract set mapped into the two element set by φ

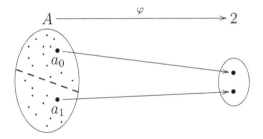

then A is divided in 2 parts. Conversely, any mutually exclusive and jointly exhaustive division of A into two parts arises from a mapping to the two-element set. (These are called either *indicator functions* or *characteristic functions* in probability theory, combinatorics, and other subjects.)

The φ is the indicator or characteristic function of the whole partition in two parts and would be determined by such a pattern. In our problem there is no specific partition. We merely want to show that a partition *can be chosen* with the two elements in different parts. This is a very weak condition on φ, so there are many such mappings; in the extreme we could imagine that only a_0 goes to one value, or in the opposite extreme that only a_1 goes to the other value. Both are extreme concepts of φ, and both would satisfy the requirement, namely, that the composites are still different. We can make that more specific: We insist that one of the composites comes out to be the element 0 and the other one the element 1, where we have structured the abstract set 2 by considering both of its elements as being listed:

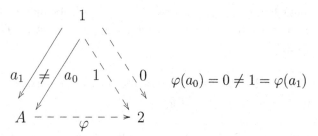

$$\varphi(a_0) = 0 \neq 1 = \varphi(a_1)$$

We can simply insist that the indexing is preserved by φ. Among these maps there is a sort of lowest possible choice for φ:

$$\underline{\varphi}(a) = \left\{ \begin{array}{ll} 1 & \text{if } a = a_1 \\ 0 & \text{if } a \neq a_1 \end{array} \right\}$$

which is one possibility, and a highest possible choice for φ,

$$\overline{\varphi}(a) = \left\{ \begin{array}{ll} 1 & \text{if } a \neq a_0 \\ 0 & \text{if } a = a_0 \end{array} \right\}$$

at the other extreme.

Both of these extreme cases of φ separate a_0, a_1. There will be many partitions φ in between these two that will also separate a_0, a_1, but since there exists at least one, we have convinced ourselves that 2 is a coseparator.

How many properties are there exactly?

Exercise 1.23
Consider a two-element set and a three-element set:

Name the elements of A by a_0 and a_1 and the elements of B by b_0, b_1, b_2. Draw the cographs of all mappings $f : A \longrightarrow B$ and all $g : B \longrightarrow A$. How many mappings are there from $A \longrightarrow B$? (We should find that there are nine.) How many from $B \longrightarrow A$? (The answer is eight $= 2^3$.) Notice that if each cograph were reduced to a dot we would obtain abstract sets of mappings with nine and eight elements; these abstract sets can in turn be used to paramaterize the actual mappings (see Chapters 5 and 7). ◊

Exercise 1.24
What about mappings from a five-element set into a two-element set? How many mappings are there from

Answer: $2^5 = 32$ mappings ◊

In general, we can say that the number of (two-valued) properties on a set with n elements is 2^n. One of the ideas of Cantor was to give meaning to this statement even when n is not finite, as we shall see in more detail later.

The coseparating property of 2 is phrased in such a way that it is really just a **dual** of the separating property that 1 has, i.e. there are enough elements to discriminate properties just as there are enough properties to discriminate elements.

Consider the various elements in, and the various 2-valued properties on the set A:

$$1 \; \overset{\longrightarrow}{\underset{a's}{\vdots}} \; A \; \overset{\longrightarrow}{\underset{\varphi's}{\vdots}} \; 2$$

We test whether a_0, a_1 are different by using such properties as follows:

$$a_0 \neq a_1 \Rightarrow \exists \varphi \text{ such that } \varphi a_0 \neq \varphi a_1$$

For each pair we find a φ; for another pair we might need another φ. The dual statement

$$\varphi_1 \neq \varphi_2 \Rightarrow \exists a \text{ at which } \varphi_1 a \neq \varphi_2 a$$

that there are enough elements, distinguishing between properties with any kind V of value specializes to a statement about two-valued properties: $\varphi_1 \neq \varphi_2$ means that two *mappings* are different, but for mappings to be different there must be some element a at which their values are different. (Again we warn the reader that in categories of less abstract sets, these separating and coseparating roles will be taken by families of objects richer than 2 and 1.)

Recall that the map $A : A \longrightarrow 1$ is unique. The statement "f is constant" means that there exists a single element of B such that f is the composite: the single element following the unique map, i.e.

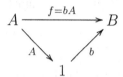

Definition 1.25: *An arrow $A \xrightarrow{f} B$ is* **constant** *if and only if it factors through 1; i.e. there is an element $b : 1 \longrightarrow B$ of B such that $f = bA : A \longrightarrow 1 \longrightarrow B$.*

In particular, if we form the composites of the unique map to 1 with the two elements of 2, we get two constant maps $2 \longrightarrow 2$:

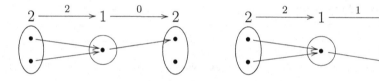

We will find various circumstances in which it is useful to look at the system of self-maps of 2 in its own right, not just as part of the category of abstract sets but as a scheme for picking out another, more interesting kind of set called reversible graph (see Section 10.3). For this and other reasons it is important to know how these four maps compose. We can write down the "multiplication table" for this four-map system. The self-maps of any given object can always be composed, and having names for them (four in this case), we can show, in the multiplication table, what is the name of the composite of any two of them taken in a given order. Here is the resulting table:

1_2	0	1	τ
0	0	0	0
1	1	1	1
τ	1	0	1_2

The composing of constants (which comes up in computing this table) works in general as follows:

If f is constant with constant value b, then for any g, gf is constant but with constant value gb

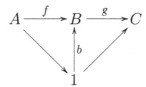

Similarly, if f is constant, then for any mapping a, fa is also constant with the same constant value as f:

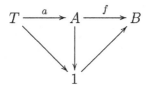

Exercise 1.26
Verify the composition table for endomappings of 2 by considering all four possibilities for each of the two arrows in $2 \longrightarrow 2 \longrightarrow 2$ and computing the composite.

$\Diamond$

Exercise 1.27
Show how this four-element system acts as "logical operators" on the properties $A \longrightarrow 2$ of (elements of) any A but also acts in another way on the lists $2 \longrightarrow A$ (or ordered pairs) of elements of A.

Hint: Recall the discussion of tennis players in Section 1.5. $\Diamond$

1.8 Additional Exercises

Exercise 1.28
Uniqueness properties:

(a) Show that identity arrows in a category are unique, i.e. if $1_A : A \longrightarrow A$ and $1'_A : A \longrightarrow A$ both satisfy the equations for identity arrows, then $1_A = 1'_A$.

(b) Show that the set 1 is unique "up to unique isomorphism," (i.e. if $1'$ also satisfies the axiom of the terminal set then there are unique mappings $f : 1 \longrightarrow 1'$ and $g : 1' \longrightarrow 1$) and that $gf = 1_1$ and $fg = 1_{1'}$.

(c) Similarly, show that the empty set 0 is unique up to unique isomorphism.

Exercise 1.29

(a) Show that if $f : A \longrightarrow B$ and $g : B \longrightarrow C$ are injective mappings, then so is gf, i.e. "composites of injectives are injective".

(b) Show that composites of surjectives are surjective.

Exercise 1.30

Categories of structures:

(a) Show that (finite-dimensional) vector spaces as objects and linear transformations as arrows form a category with composition of arrows defined by composition of mappings. Show that a one-dimensional space S is a separator; is it a coseparator? Is the terminal object a separator?

(b) Show that groups as objects and group homomorphisms as arrows form a category with composition of arrows defined by composition of mappings. Show that the additive group of integers is a separator. (Look up the definition of "group" if necessary. We will discuss groups further in Section 10.1.)

(c) Show that partially ordered sets as objects and monotone (= order-preserving) mappings as arrows form a category with composition of arrows defined by composition of mappings. Is the terminal object a separator in this category? Is there a coseparator? (Look up the definition of "partially ordered set" if necessary. We will discuss partially ordered sets further in Section 10.1.)

(d) A finite state machine with two inputs α and β is a finite set Q of "states" and a pair of endomaps of Q corresponding to α and β that effect the elementary state transitions of which the machine is capable. Given another such machine with state set Q', a homomorphism is a mapping $\varphi : Q \longrightarrow Q'$ of state sets preserving the transitions in the sense that

$$\varphi\alpha = \alpha'\varphi \text{ and } \varphi\beta = \beta'\varphi$$

where we have denoted by α' and β' the corresponding state transitions of the second machine. Define composition of homomorphisms and show that a category results.

Exercise 1.31

Categories as structures:

(a) Let V be a given vector space. Show that the following data define a category $\mathbf{V}$:

$\mathbf{V}$ has just one object called $*$;
the arrows of $\mathbf{V}$ from $*$ to $*$ are the vectors v in V;
the identity arrow for $*$ is the zero vector;
the composite of vectors v and v' is their sum.

Hint: With the data given, verify the equations of Definition 1.13.

(b) Let X be a given partially ordered set (with partial order $\leq$). Show that the following data define a category $\mathbf{X}$:

the objects of **X** are the elements of X;

for elements x, x' in X there is an arrow from x to x' exactly when $x \leq x'$.

(It follows that there is *at most one* arrow from x to x'.)

Hint: Here we did not specify the composition or identity arrows, but there is no choice about how to define them. Why?

Exercise 1.32

(a) (Dual categories) There are many methods of constructing new categories from known categories. An important example is the **dual** or **opposite** category of a category. Let $\mathcal{C}$ be a category. The dual category $\mathcal{C}^{\mathrm{op}}$ has the same objects as $\mathcal{C}$, but for objects A and B the arrows from A to B in $\mathcal{C}^{\mathrm{op}}$ are exactly the arrows from B to A in $\mathcal{C}$. Show how to define composition and identities for $\mathcal{C}^{\mathrm{op}}$ to make it a category.

(b) (Slice categories) Another important construction is the "slice category". Let $\mathcal{C}$ be a category and X an object of $\mathcal{C}$. The slice category of $\mathcal{C}$ by X, denoted $\mathcal{C}/X$, has objects the *arrows of $\mathcal{C}$ with codomain X*. Let $f : A \longrightarrow X$ and $g : B \longrightarrow X$ be objects of $\mathcal{C}/X$. The arrows of $\mathcal{C}/X$ from f to g are arrows h of $\mathcal{C}$ such that $f = gh$, i.e. they are the same thing as commutative triangles from f to g. Composition and identities are inherited from $\mathcal{C}$. Verify that $\mathcal{C}/X$ is a category.

(c) (Pointed sets) If we define objects to be pairs consisting of a set A and an element $1 \xrightarrow{a} A$ of A, and arrows from A, a to B, b to be mappings $A \xrightarrow{f} B$ such that $fa = b$, then we obtain a category denoted by $1/\mathcal{S}$. Verify that $1/\mathcal{S}$ is a category. Does $1/\mathcal{S}$ have an initial object? Does it have a terminal object? Does the terminal object separate mappings in $1/\mathcal{S}$? (This is a special case of a dual notion to slice category: For $\mathcal{C}$ a category and X an object of $\mathcal{C}$, describe the category $X/\mathcal{C}$.)

2

Sums, Monomorphisms, and Parts

2.1 Sum as a Universal Property

Our basic definitions of the fundamental operations

ADDITION   MULTIPLICATION   EXPONENTIATION

are all **universal mapping properties;** that is, we take as the defining condition on such a construction essentially "what it is good for".

Let us consider sums. Our naive picture of the sum is that the sum of two sets contains two parts, that the two parts are the same size as the given sets, that the sum does not have anything in it except those two parts, and that the two parts do not overlap. These ideas can be expressed more briefly as follows:

The two parts are exhaustive and mutually exclusive.

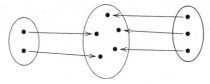

The sum as defined by the universal mapping property will, in particular, have two mutually disjoint parts equivalent to two given sets and that together exhaust the whole sum. However, it is not satisfactory to take these conditions as a definition for at least two different reasons:

(1) There are categories other than abstract sets; for objects in these categories there is also a notion of sum but one in which the two parts may interact. The description in terms of nonoverlapping would be incorrect in such a case (see Exercise 2.40). It is better to use the same form of the definition so that we see the similarity more transparently.

(2) The second reason is related to the gap between the naive picture and the formalized axiomatics. We always have some kind of naive picture for the

things we are trying to describe by mathematical formalism. But the naive picture usually does not match up perfectly with the precise formalism. There is a gap between the two. We try to get them as close together as we can. That is one of the struggles that moves mathematics forward. In particular, the words "mutually exclusive and exhaustive pair of parts" describe formally much of the naive picture of a sum. However, if we took that as the formalized axiom, it would not formally imply the needed stronger conclusion that we can always do certain things with the sum. We take those certain uses as the axiom.

First, let us consider the particular case of the sum of two copies of a one-element set.

Any two-element set will have two arrows from 1 (elements in the narrow sense); we could give them names 0 and 1.

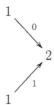

That the set 2 has two parts which are disjoint and exhaustive means that these two arrows are different and that there are no other arrows $1 \longrightarrow 2$. If we tried to take that as our definition, we would meet a difficulty with other things we want to do with 2: Namely, what is a mapping from 2 into another set B? It should just mean to give a pair of elements of B because such a single mapping has to have a value at one element and also a value at the other element.

Obviously *if we had* a mapping $2 \xrightarrow{f} B$, we could compose it with the two elements of 2, and we would obtain two elements of B.

Conversely, given two elements of B, say b_0 and b_1, then the naive picture of 2 and the "arbitrariness" of mappings suggest that there exists a unique mapping $f : 2 \longrightarrow B$ whose values are the given elements. We are led to define this particular sum as follows:

Definition 2.1: *The statement*

$$2 = 1 + 1$$

means that

there are given mappings $1 \xrightarrow{\ 0\ } 2 \xleftarrow{\ 1\ } 1$ *such that*

$\forall B, 1 \xrightarrow{b_0} B, 1 \xrightarrow{b_1} B \ \exists! f \ [f(0) = b_0 \ and \ f(1) = b_1]$

as in

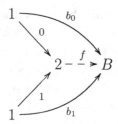

The uniqueness means that if we had two mappings f, g, both with the specified values, then they would be equal. We already introduced the principle that there are enough mappings from 1 to distinguish maps. From that the uniqueness part of the condition on f would follow if there were only the two maps from 1 to 2; that is true for the category of abstract sets (but not for all categories of interest). Concerning the existence of f, the naive idea that arbitrary mappings exist is a *guide*, but we want to prove all of the existence theorems from just a few; this is one of the few that we assume. For any set B, to map 2 into B it suffices to specify two elements; then there will be one and only one f. The exhaustiveness and disjointness of elements will follow (with the help of the other axioms). Knowing how a given object maps to all objects B in its category determines indirectly how each object maps into it, but some work will be required to realize that determination, even for the object 2.

Recall the contrast of listings versus properties (see Chapter 1). A sum of 1's (such as 2) is clearly well-adapted to a simple listing of elements of an arbitrary object B. In the category S of abstract sets (and in a few others), 2 also serves well as a codomain for properties.

It is really no more difficult to describe the sum of any two sets; but the sum of two sets is not merely a third set; rather, it is two maps as follows:

Definition 2.2: *A **sum** of the two sets A_0 and A_1 is a set A together with given mappings $A_0 \xrightarrow{i_0} A \xleftarrow{i_1} A_1$ such that*

$$\forall B, A_0 \xrightarrow{f_0} B, A_1 \xrightarrow{f_1} B \;\; \exists! f \; [f i_0 = f_0 \text{ and } f i_1 = f_1]$$

as in the commutative diagram

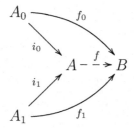

We denote the unique f defined by the mappings f_0 and f_1 by

$$f = \left\{ \begin{array}{l} f_0 \\ f_1 \end{array} \right.$$

We abbreviate by writing $A = A_0 + A_1$ with the mappings i_0 and i_1 understood.

Thus, given any two sets A_0 and A_1, to make A the sum of the two, we must give i_0 and i_1 with the universal mapping property presented at the beginning of this Section. For any B at all, if we have a map $A \longrightarrow B$, then of course, by composing, we get two maps $A_0 \longrightarrow B$ and $A_1 \longrightarrow B$; but the sum injections i_0 and i_1 are special so that conversely, given any f_0 defined on one part A_0 and any f_1 defined on the other part A_1, there is exactly one f defined on the whole that composes with the i_0, i_1 to give the original f_0, f_1.

AXIOM: BINARY SUMS
Any two sets A_0, A_1 have a sum $A_0 \overset{i_0}{\longrightarrow} A \overset{i_1}{\longleftarrow} A_1$.

Exercise 2.3
Given a set A_0 with exactly two elements and a set A_1 with exactly three elements, specify a set A and two injection maps for which you can prove the universal mapping property of a sum of A_0 and A_1. That is, given any f_0 and f_1, show how to determine an appropriate f.   ◊

EXAMPLE: COGRAPHS
We have already seen an important example of a mapping from a sum. The **cograph** of a mapping, which we described with a picture in Section 1.1, is actually an instance. Indeed, suppose that $f : A \longrightarrow B$ is any mapping whatsoever. We can define a mapping $c : A + B \longrightarrow B$ by $c = \left\{ \begin{array}{l} f \\ 1_B \end{array} \right.$. That is, the following is a commutative diagram:

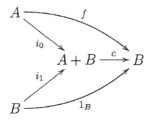

The cograph of a mapping is illustrated by its internal diagram (which we have already used informally many times!) by drawing the sets A and B disjointly and showing a linking of an element a of A to an element b of B whenever $f(a) = b$;

thus, two elements in $A + B$ are linked in the internal diagram if and only if they are merged by the cograph mapping c.

Exercise 2.4

Illustrate the example from Exercise 2.3 by an internal cograph picture.   ◊

Coming back to Definition 2.2, for any element $1 \xrightarrow{a} A$, we have

$$(*) \quad \begin{aligned} fa &= f_0 a_0 \quad \text{if } i_0 a_0 = a \quad \text{for some element } a_0 \text{ of } A_0 \\ fa &= f_1 a_1 \quad \text{if } i_1 a_1 = a \quad \text{for some element } a_1 \text{ of } A_1 \end{aligned}$$

Naively, there could not be two different maps f that satisfy the condition $(*)$, for A is exhausted by the two parts; if there were an element a that could not be expressed in either of these two forms, then we could imagine that there could be maps f for which the equations $(*)$ would be true and yet f could have various values at such a. In other words the exhaustiveness of the two parts corresponds naively to the uniqueness of the f. On the other hand, the disjointness has something to do with the existence of the f. What would happen if we had an element a that could be fed through both i_0 and i_1 (i.e., the two parts overlap, at least on that point)?

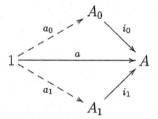

Then we could imagine that f_0 and f_1 take different values at a_0 and a_1; therefore, there could not be any f at all because even fa would not be well-defined. The naive content of the disjointness and exhaustiveness corresponds to the existence and uniqueness of such maps f.

We will write $A_0 + A_1 = A$ as a kind of shorthand; it means that whenever we need to apply the universal property we are given the i_0 and the i_1 required. The universal mapping property has as a by-product that the "size" of A is determined by the sizes of A_0 and A_1, but it has many more detailed uses than the mere "size" statement.

How does the general sum relate to 2? If we are given a decomposition of any set A as a sum of two other sets A_0 and A_1, we will automatically have a mapping from A to 2, namely, the mapping that takes the two values on those two respective parts. That follows from our definition, for we could apply the universal mapping property by taking $B = 2$ and consider the two elements (one called 0 and the other called 1) of 2; each of those can be preceded by the unique map that always exists from any set to the one-element set; i.e. we can take f_0 and f_1 to be, respectively,

A_0's unique map to 1 followed by the element 0 of 2

A_1's unique map to 1 followed by the other element 1 of 2

Now we have two maps, f_0 and f_1 going from A_0 and A_1 into the set 2, and so there is a unique f whose domain is the sum, whose codomain is 2, and such that these composites are equal to the respective constant mappings just constructed, as in the following commutative diagram:

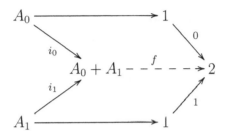

On the part i_0 of the sum, this mapping f constantly takes the value 0; the other composite is constantly the other element 1 of 2.

Conversely, any mapping $A \xrightarrow{f} 2$ really is of the preceding sort because, if we are given the f, then we can define two parts of which A is the sum. We can imagine that each part is the one whose members are all the elements for which f has a fixed value. That is, we may construct mappings into 2 by other means and then deduce the corresponding sum decomposition.

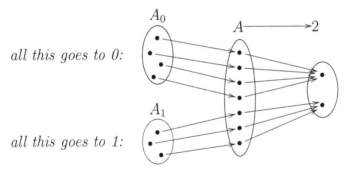

Anytime we have this situation, we may say A has been divided into two parts. But what is a part and what is membership? We will discuss these concepts in the next section.

Exercise 2.5

Define explicitly what should be meant by a sum of three given sets. If V is any fixed set with exactly three elements, show that any mapping $A \longrightarrow V$ comes from a unique sum decomposition of A into three parts.   ◊

2.2 Monomorphisms and Parts

Monomapping, or *monomorphism*, is a stronger version of *injective*, which we will have to use to define *part* in general categories of sets and for other purposes also:

Definition 2.6: *An arrow* $S \xrightarrow{i} A$ *is a* **monomapping** *or* **monomorphism** *if it has the cancellation property with respect to composition on its right, i.e.,*

$$\forall T \; \forall s_1, s_2 : T \longrightarrow S \; [i s_1 = i s_2 \Rightarrow s_1 = s_2]$$

so whenever we have the configuration

$$T \underset{s_2}{\overset{s_1}{\rightrightarrows}} S \xrightarrow{i} A$$

and $i s_1 = i s_2$, *then* $s_1 = s_2$.

Equivalently,

$$\forall T \; \forall x \; [T \xrightarrow{x} A \Rightarrow \exists \text{ at most one } s \; [i s = x]]$$

The expression above is called a cancellation property because, given an equation involving composition with i on the left, we can cancel the i. Since s_1 and s_2 are generalized elements, the property says that i is injective even on generalized elements. The difference between *monomorphic* and *injective* is that, for the "mono" property, we require the cancellation for all T. This does not matter in the case of abstract sets, where cancellation with the general T or with just $T = 1$ means the same thing (see Exercise 2.7). That "mono" implies *injective* is tautologous because a general statement always implies any of its special cases. The converse statement is *not* tautologous; it depends on the existence of sufficiently many elements.

Exercise 2.7
Using the axiom that 1 separates mappings, show that

$$\text{injective} \Rightarrow \text{monomapping}$$

in the category of sets S.   ◊

Exercise 2.8
Show that if i and j are composable monomorphisms, then ji is a monomorphism, i.e. a composite of monomorphisms is a monomorphism. (Recall that injective mappings also compose.)   ◊

Exercise 2.9

Show that any mapping $1 \xrightarrow{a} A$ whose domain is 1 is necessarily a monomapping.
◊

Exercise 2.10

Show that if U has the property that the unique $U \longrightarrow 1$ is a monomapping, then any mapping $U \xrightarrow{a} A$ with domain U is necessarily a monomapping. ◊

The origin of these words (including the term *epimorphic*, seen later to be related to surjective) is in ancient Mediterranean languages as follows:

GREEK:	LATIN:
monomorphic	injective
epimorphic	surjective

We exploit the differences in the words to get slightly different meanings. The difference between *injection* and *monomorphism* is not too great in practice, whereas the difference between *surjection* and *epimorphism* (the dual of monomorphism, which we will define later) is greater. What we mean by monomapping or injective mapping is one that does not identify (in the sense of "make equal") any elements that were previously distinct.

We are now ready to define *part*.

Definition 2.11: *A* **part** *of set A is any mapping i for which*

(1) *the codomain of i is A, and*
(2) *i is a monomapping.*

Any part of A has some domain, say S, but the property (2) means that the part keeps the elements distinct: if we have $is_1 = is_2$, we can conclude $s_1 = s_2$. In particular, if we return to the definition of *sum*, the injections into a sum have to be monomappings.

Exercise 2.12

Prove on the basis of the axioms for S so far introduced that if $A_0 \xrightarrow{i_0} A \xleftarrow{i_1} A_1$ is a sum in S, then i_0 is a monomapping. ◊

Here is an example of the notion of part: There is the set A of chairs in a room and the set S of the students in the room. The mapping i assigns to every student the chair in which she or he sits. That is an injective mapping. Why? Because we have no situation of two students sitting in one seat. It is a monomorphism and therefore defines a part of the set of all seats. Students are not chairs – we have to have a given mapping that *specifies* the way in which we have a part. The account of "part of" given in most books would make no sense for abstract sets: It is not that every element of S is an element of A! In our concrete example, the latter would mean that every student "is" a chair. But i is indeed a part in the sense that we want to use the term in mathematics; it does define a condition on chairs – to be a member of the part means to be an occupied chair.

2.3 Inclusion and Membership

To express the relationship between parts that arises when they can be compared, we have

Definition 2.13: *The part* i *is* **included in** *the part* j *(or there is an* **inclusion** *from* i *to* j*) means that* i *and* j *are parts of the same* A *and that*

$$\exists k[i = jk]$$

We write $i \subseteq_A j$ *in this case and we sometimes omit the subscript* A *if the codomain* *("universe of discourse") is obvious.*

Exercise 2.14

Show that if i is a monomorphism and $i = fk$, then k is a monomorphism. Give an example to show that f need not be injective.   ◊

Thus, the k whose existence is required in the definition is always a monomapping, and so k is a part of the domain V of j.

Definition 2.15: *The (generalized) element* a *of* A *is a* **member** *of the part* i *of* A, *denoted* $a \in_A i$, *means that* i *is a monomapping and that*

$$\exists \bar{a}[a = i\bar{a}]$$

Notice that we give the definition of membership for *generalized* elements. In the special case $T = 1$ we are stating that an element a of A is a member of the part

i of A if there is an element $\bar{a}$ of the domain U of i that the part i interprets as a. (Note that there can be at most one such $\bar{a}$.)

Proposition 2.16: *For any a, i and j (the "universe" A being understood),*

$$a \in i \text{ and } i \subseteq j \Rightarrow a \in j$$

Proof: The hypothesis gives us $\bar{a}$ and k for which $a = i\bar{a}$ and $i = jk$. To establish the conclusion we would need $\bar{\bar{a}}$ with $a = j\,\bar{\bar{a}}$. The only map at hand that even has the right domain and codomain as such an $\bar{\bar{a}}$ is $k\bar{a}$; so define $\bar{\bar{a}} = k\bar{a}$ as in the diagram

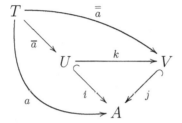

and see whether it satisfies the correct equation:

$$j\,\bar{\bar{a}} = j(k\bar{a})$$
$$= (jk)\bar{a}$$
$$= i\bar{a}$$
$$= a$$

Hence, $a = j\,\bar{\bar{a}}$ as required to prove that $a \in j$. ∎

Exercise 2.17
If $a \in i$, then for all t, $at \in i$. ◊

In order to follow the next construction it will be useful to make explicit the logical rule of inference for implication–introduction.

If X, Y, Z are statements, then

$$(X \text{ \& } Y) \Rightarrow Z \ \ holds$$
$$if \ and \ only \ if$$
$$Y \Rightarrow (X \Rightarrow Z) \ \ holds$$

(For more details about the symbols, see Appendix C.2 and Appendix A.)

For example, if one takes X to be the statement $a \in j$, Y to be $i \subseteq j$, and Z to be $a \in j$, the proposition we have just proved is the statement with & in the hypothesis. Thus, we could equally well say that we have proved the statement with the $\Rightarrow$ inside the conclusion:

$$i \subseteq j \Rightarrow [a \in i \Rightarrow a \in j]$$

Note that the variable a now appears in the conclusion but not in the hypothesis.

We also need to make explicit the

logical rule of inference for $\forall$-introduction.

If Y and W are statements in which W depends on a variable a but Y does not, then

$$Y \Rightarrow W \quad holds$$

if and only if

$$Y \Rightarrow \forall a[W] \quad holds$$

Thus, our proposition above is further equivalent to

$$i \subseteq_A j \Rightarrow \forall a[a \in i \Rightarrow a \in j]$$

where the statement on the right has the universal quantifier $\forall a$ ranging over all $T \xrightarrow{a} A$ for all sets T. The converse of (this form of) the proposition

$$\forall a[a \in i \Rightarrow a \in j] \Rightarrow i \subseteq_A j$$

is also of interest. This converse is trivially true if the domain of a is arbitrary since, for the special case $a = i, \bar{a} = 1_U$ proves $i \in i$; hence, if the hypothesis $\forall a[\ldots \Rightarrow \ldots]$ is true, we get $\bar{\bar{a}}$ proving that $i \in j$, but that is the same as a proof k that $i \subseteq j$. However, the converse becomes more interesting if we restrict the range of the a's for which we assume the hypothesis, for then it could be a nontrivial property of our category that the conclusion $i \subseteq j$ nonetheless still followed from this weaker hypothesis. Something of this sort will be true of categories of variable sets, where it will be possible to restrict a to the indispensable domains of variation of the sets. In the case of abstract (= constant) sets there is no substantial domain of variation, and so we can restrict the domain of a to be 1. That is, the category of abstract sets satisfies the strengthened converse proposition (a weakened hypothesis)

$$\forall a[1 \xrightarrow{a} A \Rightarrow (a \in i \Rightarrow a \in j)] \Rightarrow i \subseteq j$$

Essentially, this follows from the idea of the arbitrariness of mappings: Note that what is wanted in the conclusion is a single actual map k for which $i = jk$, yet the hypothesis only says that

$$\forall 1 \xrightarrow{x} U \; [\exists 1 \xrightarrow{y} V \quad jy = ix]$$

(Here the a is the common value $jy = a = ix$). Since j is injective, the $\exists y$ can be strengthened to $\exists!y$.

Thus, from the hypothesis we get

$$\forall x\ \exists!y\ [jy = ix]$$

The idea that mappings are arbitrary means that, whenever $\forall x \exists!y[\ldots]$ holds, then there does exist a mapping k – in this case one for which

$$\forall x[j(kx) = ix]$$

Since there are "enough" $1 \xrightarrow{x} U$, we conclude that $jk = i$, as required.

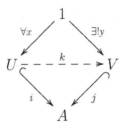

We will soon be able to prove this strengthened converse on a more explicit basis that does not depend on the vague "arbitrariness of mappings" principle.

Historically many mathematicians such as Dedekind and Banach used the same symbol $\subseteq$ for both membership and inclusion, and indeed it is quite reasonable to define x **belongs to** y for any two maps with codomain A and even to use the same symbol for it

$$x \subseteq y \text{ iff } \exists z[x = yz]$$

where iff is the standard abbreviation for "if and only if".

If y is not monomorphic, then such a z may not be unique; such z are often called "liftings of x along y". In the formalized set theory following Peano, the membership symbol was given a completely different meaning, but mathematical use of it has been as we defined; thus, we feel justified in using the membership symbol for this special case of the "belonging to" relation even though it is not strictly needed. The notion obtained by reversing the arrows

$$\exists g[f = gh]$$

may be called f **is determined by** h and is at least as important in mathematics and its applications. Here the g, which shows a way of determining f values from h values, is often called an "extension of f along h"; such a g is only guaranteed to be unique if h is "epimorphic" (the dual cancellation property). Epimorphic maps with a given *domain A* are *partitions* (as opposed to parts) of A (see Section 4.4); again the special case where both f and h are partitions rates special attention: in that case the determining g is often called a "coarsening" of partitions (rather than an including of parts).

Exercise 2.18

If f is determined by h, then any further φf is also determined by h.

General belonging and determining often reduce to the special cases of belonging to parts (i.e., membership) and, respectively, of determining by partitions. This happens because frequently a map y can be expressed as $y = y_0 y_1$, where y_0 is "mono" and y_1 is "epi"; then, if x belongs to y, it follows that x is a member of the part y_0 of A:

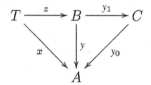

If z proves that x belongs to y, then $y_1 z$ proves the stated membership; conversely, if $x \in y_0$, then in a sense x "locally" belongs to y.   ◊

Exercise 2.19

Formulate the dual construction, showing that if f is determined by h and if h has a "mono/epi" factorization $h = h_0 h_1$, then f is an invariant of the partition h_1 of the domain. (In suitable categories the converse will say that if f is an invariant of h_1, then some "expansion" of f is determined by h; the phrase "invariant of" is frequently used in this context and is dual to "member of," and we have used the term "expansion" dually to a common use of the term "covering" to explain "holding locally".)   ◊

2.4 Characteristic Functions

Given a chosen set V thought of as "truth values" and a chosen constant element $1 \xrightarrow{v_1} V$ thought of as "true," we can define a relationship as follows:

Definition 2.20: *For any set A, mapping i with codomain A, and mapping $A \xrightarrow{\varphi} V$, φ is an* **indicator** *or* **characteristic function** *of i if and only if*

(1) *i is monomorphic and*
(2) *for any T and any $T \xrightarrow{a} A$*

$$a \in_A i \iff \varphi a = v_1 T$$

(For any $1 \xrightarrow{c} C$, cT denotes the constant map

$$T \xrightarrow{T} 1 \xrightarrow{c} C$$

with domain T and value c).

If we refer to the element v_1 as "true," then a is a member of i if and only if φa is "true" since that is true for all a:

$$\varphi i = v_1 U \quad \text{where } U = dom(i)$$

moreover for any a, if $\varphi a = v_1 T$, then there is $\bar{a}$ for which $a = i \bar{a}$:

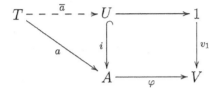

Exercise 2.21
For example, if $A = V$ and $i = v_1$, then an indicator of v_1 as required above is $\varphi = 1_V$.   ◊

The exercise shows that (1) v_1 is monomorphic, and (2) for any T and any $T \xrightarrow{a} V$

$$a \in_V v_1 \iff 1_V a = v_1 T$$

The idea of the indicator, then, is that the general membership relation can be represented by this particular case. An important fact about the category of abstract sets is that for $V = 2$ and the chosen element $1 \xrightarrow{1} 2$ we have *unique* characteristic functions for all parts of all sets. That is,

AXIOM: MEMBERSHIP REPRESENTATION VIA TRUTH VALUES

(1) *Any $A \xrightarrow{\varphi} 2$ is the characteristic function of a part of A.*
(2) *Any part of A has a unique 2-valued characteristic function.*

For categories of cohesive and variable sets there will still be a $V \neq 2$ satisfying properties (1) and (2), but that the set 2 has these two properties is special to the category $\mathcal{S}$ and those special categories of variable sets called Boolean toposes.

The preceding principle will have a great many uses. It enables us, for example, to count parts by counting 2-valued functions instead. Recall that counting requires a clear notion for distinguishing the things being counted.

Proposition 2.22: *The following are equivalent for parts* $U \xrightarrow{i} A$ *and* $V \xrightarrow{j} A$ *of a set A:*

(1) $i \subseteq_A j$ & $j \subseteq_A i$,
(2) $\exists h, k[j = ih$ & $i = jk$ & $hk = 1_U$ & $kh = 1_V]$, *and*
(3) $\varphi = \psi$, *where these are the characteristic functions of i and j, respectively.*

Exercise 2.23

Prove the proposition, i.e. show that each of (1), (2), and (3) implies the others.

Hint: (2) obviously implies (1), but to see the converse remember that i and j are parts. $\diamond$

Definition 2.24: *Parts i and j of a set A are **equivalent**, which we write $i \equiv_A j$ if and only if $i \subseteq_A j$ & $j \subseteq_A i$.*

Thus, characteristic functions do not distinguish equivalent parts, and parts are distinguished for counting purposes only if they are not equivalent. Indeed, we will sometimes abuse language by writing equivalent parts as equal. However, even if two parts have the same number of elements in the sense we will describe soon in Section 3.2, they need not be equivalent.

Exercise 2.25

If i and j are parts of A and their mere domains are isomorphic, meaning that there are h, k satisfying $hk = 1_U$ & $kh = 1_V$, then neither part need be included in the other. However, if k also proves that $i \subseteq_A j$, then it follows that $i \equiv_A j$ as parts. $\diamond$

2.5 Inverse Image of a Part

A frequent use of parts is in *restriction* of mappings.

Definition 2.26: *If $X \xrightarrow{f} Y$ and if i is a part of X, then the composite fi*

$$U \xrightarrow{i} X \xrightarrow{f} Y$$

*is called the **restriction** of f to the part i (often denoted $f|i$ or $f|_i$).*

Exercise 2.27

Give an example of a surjective f and of a part of its domain such that the restriction is *not* surjective. Give an example of f that is not injective but such that a restriction of f *is* injective. $\diamond$

For example, if f describes the temporal variation of temperature throughout a year and if j is the subzero part of the temperature values, then there can be a certain week i in January such that $fi \in j$.

Exercise 2.28
Monomaps are only left-cancellable and are not usually right-cancellable. Give an example of distinct f, g with equal restrictions to the same part.   ◊

An operation of fundamental importance in geometry, analysis, logic, and set theory is that of forming the

INVERSE IMAGE OF A PART ALONG A MAP

which also goes under names like "substitution" or "pullback" in certain contexts. Thus, if we are given an arbitrary map f from X to Y and an arbitrary part $V \overset{i}{\hookrightarrow} Y$ of the codomain, then there is a part $U \overset{i}{\hookrightarrow} X$ of the domain such that

(0) for all $T \overset{x}{\longrightarrow} X$

$$x \in i \Longleftrightarrow fx \in j$$

This is easily seen to imply that
(1) there is a commutative square

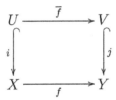

and that, moreover,
(2) any commutative square

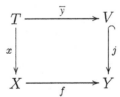

can be expressed in terms of (1) by means of a common $\bar{x}$:

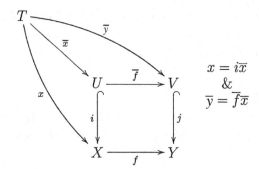

In other words, f restricted to the part i has its image in j, and i is the largest part of X with that property.

(3) The $\bar{x}$ is unique (because we have assumed that i is a monomapping).

Exercise 2.29

(a) Show that (0) $\Rightarrow$ (1), (2), and (3).

(b) Show that (1), (2), (3), and j is mono $\Rightarrow$ i is mono.

(c) Show that (1), (2), and (3) $\Rightarrow$ (0).

(d) Show that for given f, j, if both i_1, i_2 satisfy (0), then $i_1 \equiv_X i_2$ as parts of X.

$\Diamond$

The part i described above is called the **inverse image** of j along f. The operation of inverse image has the following contravariant functoriality property:

Exercise 2.30

If j is the inverse image of k along g and if i is the inverse image of j along f, then i is the inverse image of k along gf.

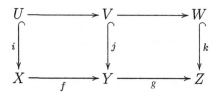

$\Diamond$

When there is a truth-value object V (such as the set 2 in the case of abstract sets), i.e. an object "classifying" all parts by means of unique characteristic functions, then inverse image is so intimately related to composing that the functoriality above translates into a case of associativity (see Exercise 2.34).

Proposition 2.31: $Y \xrightarrow{\psi} V$ *is the characteristic function of a part j if and only if j is the inverse image along ψ of the one-element part $1 \xrightarrow{v_1} V$ of V.*

Exercise 2.32
Prove the Proposition 2.31.

Hint: Apply the definition of indicator. ◇

Theorem 2.33: *If ψ is the characteristic function of a part j, then for any f, ψf is the characteristic function of the inverse image of j along f.*

Proof: Apply the functoriality of inverse image to the special case indicated.

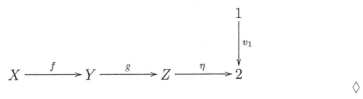

■

Exercise 2.34
Interpret $\eta(gf) = (\eta g)f$ in terms of the inverse images:

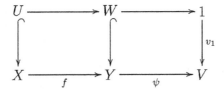

◇

An important special case of inverse image is **intersection,** which is what results from inverse image along a map that is itself a part.

Exercise 2.35
The composite of monomaps is again such. Hence, if both f, j are monomaps and i is the inverse image of j along f, then $m = fi = j\overline{f}$ is also a part, and for any y,

$$y \in m \iff [y \in j \ \& \ y \in f]$$

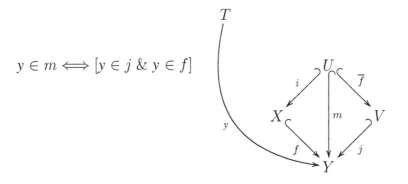

◇

Definition 2.36: *For parts i and j with the same codomain Y, a part m of Y is called their* **intersection** *if and only if for all T, for all* $T \xrightarrow{y} Y$

$$y \in m \iff [y \in i \ \& \ y \in j]$$

The m so defined is usually denoted $m = i \cap j$.

Remark 2.37: Exercise 2.35 proves that the inverse image of a part j along a monomapping f is the intersection of the two parts j and f.

Exercise 2.38
Prove that we have $j_1 \cap j_2 \equiv j_2 \cap j_1$.   ◊

Inverse images along arbitrary maps preserve intersection of parts in the following sense:

Exercise 2.39
If $X \xrightarrow{f} Y$ and if j_1, j_2 are any two parts of Y, and if i_k is the inverse image of j_k along f for $k = 1, 2$, then the inverse image of $j_1 \cap j_2$ along f is $i_1 \cap i_2$, where the latter indicates intersection of parts of X. Briefly, $f^{-1}(j_1 \cap j_2) \equiv f^{-1}(j_1) \cap f^{-1}(j_2)$, where we have exploited the essential uniqueness of the inverse image operation to justify introducing a notation $f^{-1}(\ )$ for it.   ◊

2.6 Additional Exercises

Exercise 2.40
The category of vector spaces and linear transformations has sums, but they are very different from the sums in S. Let V and V' be vector spaces.

(a) Show that the vector space $V \times V'$ of ordered pairs consisting of a vector from V and one from V' (specify how this is a vector space!) is the sum.

   Hint: The injection $V \xrightarrow{i_0} V \times V'$ sends v to $(v, 0)$.

(b) Show that monomorphisms are injective linear transformations (i.e., those with kernel 0). Thus, a subspace of a vector space determines a monomorphism to the vector space (*part* of the space).

Recall that an object 1 that satisfies the axiom of the one-element set in a category (i.e., for any object there is a unique map to 1) is called a *terminal object* of the category. Dually, an object 0 that satisfies the axiom of the empty set in a category (i.e., for any object there is a unique map from 0) is called an *initial object* of the category.

(c) Show that the zero vector space is simultaneously a terminal object and an initial object (such an object is called a **zero object**).

(d) Show that any part of a vector space has an inverse image along any linear transformation. Remember to verify that the inverse image of the mapping is a vector space.

Exercise 2.41

(a) Show that injective group homomorphisms are monomorphisms in the category of groups and group homomorphisms.

(b) Show that the one-element group is a zero object.

The category of groups also has sums, but their construction is rather complicated – they are sometimes called "free products".

(c) Show that any part (= subgroup) of a group has an inverse image along any group homomorphism.

Exercise 2.42

(a) Show that the category of partially ordered sets has sums.

   Hint: The sum partial order is on the sum of the two sets involved. Elements are related in the sum exactly if they are related in a component of the sum of sets.

(b) Show that monomorphisms in the category of partially ordered sets are injective monotone mappings.

(c) Show that the one-element set and the empty set with the only possible order relations are the terminal and initial ordered sets.

Exercise 2.43

An important case of slice categories (see Exercise 1.30(e)) is the category of X-*indexed families of abstract sets* S/X. Recall that in S/X objects are mappings with codomain X and arrows are commutative triangles. The name "family" arises as follows: For any object $A \xrightarrow{f} X$ of S/X and any element $1 \xrightarrow{x} X$ of X the inverse image of x along f is a part of A denoted A_x and is called "the fiber of A over x;" A is the "sum" of the family of all of its fibers. This is a very simple example of a variable set.

(a) Show that the category S/X has binary sums that are computed "fiberwise".

(b) Show that monomorphisms in S/X are also "fiberwise" and have characteristic morphisms taking values in the object Ω of S/X, which has each fiber equal to 2.

(c) Show that the terminal object in S/X is $X \xrightarrow{1_X} X$, and the initial object is the unique $0 \longrightarrow X$.

(d) Show that S/X has inverse images computed using the inverse images in S.

Exercise 2.44

(a) What are the parts of the terminal object in S/X (see the previous exercise)? Show a one–one correspondence between them and elements of the object Ω above (i.e., arrows from the terminal object to Ω).

(b) Show that the terminal object in S/X is **not** in general a separator. However, show that for any parallel pair of different maps in S/X there is a map from a *part* of the terminal object to the common domain that distinguishes the parallel pair.

Exercise 2.45

A special case of the preceding examples occurs when $X = 2$. Here 2 has the elements 0 and 1, and so each object of $S/2$ is just the sum of its two fibers. A related category is the "category of pairs of sets," which we denote $S \times S$. The objects of this latter category are pairs written (A_0, A_1) of objects of S. An arrow from (A_0, A_1) to (B_0, B_1) is a pair of mappings $A_0 \xrightarrow{f_0} B_0$, $A_1 \xrightarrow{f_1} B_1$.

(a) Show that $S \times S$ is a category.

(b) Show that any object (A_0, A_1) of $S \times S$ defines a unique object of $S/2$, and conversely. Show that any arrow of $S \times S$ defines a unique arrow of $S/2$, and conversely.

(c) Show that the correspondences in (b) respect identities and composition and have the property that when their actions on objects are "composed" with each other the resulting object is isomorphic to the starting object. In this situation, we say that $S \times S$ and $S/2$ are **equivalent** as categories.

Exercise 2.46

(a) Show that the category of pointed sets $1/S$ (see Exercise 1.30(f)) has sums.

   Hint: The sum of A, a and B, b is the set formed by taking the sum of the sets A and B and then merging the elements a and b in $A + B$. Thus, the sum of a two-element pointed set and three-element pointed set has how many elements?

(b) Describe monomorphisms in $1/S$.

(c) Show that $1/S$ has inverse images.

Exercise 2.47

Let **X** be the category defined by an ordered set X (see Exercise 1.31(b)).

(a) Show that objects x, x' of **X** (elements of X!) have a sum exactly when they have a *least upper bound*.

(b) Show that *any* arrow of **X** is a monomorphism.

(c) Show that a greatest element of X is a terminal object in **X**, whereas a smallest element of X is an initial object.

(d) Show that a part of x (i.e., $u \leq x$) has an inverse image along an arrow to x (i.e., $x' \leq x$) if and only if u and x' have a *greatest lower bound*.

3

Finite Inverse Limits

3.1 Retractions

We begin this chapter by observing an important condition which implies that a mapping is a monomapping, namely, that it has a *retraction.*

Definition 3.1: *For mappings r and i in any category, r is a* **retraction** *for i means that ri is an identity mapping. In the same situation we say that i is a* **section** *for r.*

To see more clearly what this says about domains and codomains, let $X \xrightarrow{i} Y$; then, we see that r must have domain Y (in order that ri be defined) and codomain X (in order that ri could possibly be an identity mapping, necessarily that on X).

$$X \underset{i}{\overset{r}{\longleftarrow \longrightarrow}} Y \qquad ri = 1_X$$

For a given mapping i, there may or may not be a retraction r for i, and there might be many. (Similarly, a mapping r need not have any sections.) But to have even one retraction is a strong condition:

Proposition 3.2: *If i has at least one retraction r, then i is a monomapping ($=$ left-cancellable mapping, which we have seen is the same as injective mapping.)*

Proof: We must show that for any x_1, x_2 as shown

$$T \underset{x_2}{\overset{x_1}{\rightrightarrows}} X \xrightarrow{i} Y$$

that $i x_1 = i x_2 \Rightarrow x_1 = x_2$.

So suppose $ix_1 = ix_2$. Then,

$$r(ix_1) = r(ix_2), \quad \text{so associativity of composition}$$
$$(ri)x_1 = (ri)x_2, \quad \text{gives but } ri \text{ being an identity}$$
$$x_1 = x_2 \qquad \text{map, this is as desired}$$

■

(Note how closely this proof parallels the sort of calculation done in elementary algebra: Suppose we know $7x_1 = 7x_2$; then,

$$\tfrac{1}{7}(7x_1) = \tfrac{1}{7}(7x_2)$$
$$(\tfrac{1}{7}7)x_1 = (\tfrac{1}{7}7)x_2$$
$$1x_1 = 1x_2$$
$$x_1 = x_2$$

The construction in our proof is analogous to *division* by 7; i.e. to *multiplication* by a multiplicative inverse for 7. The only new observation is that $\tfrac{1}{7}$ does not really need to be an inverse for 7, that is, to satisfy both $\tfrac{1}{7}7 = 1$ and $7\tfrac{1}{7} = 1$; only the first of these equations was used – it is enough to have a *left-inverse* to be able to cancel a factor on the left.)

Notice that this calculation gives more than merely knowing that i can be cancelled; having r provides a specific reason, or proof, of the left-cancellability of i.

A mapping i that has a retraction is also called a **split injection** or **split mono-mapping**. Thus,

Definition 3.3: *The arrow $i : X \longrightarrow Y$ is a* **split injection** *if and only if*

$$\exists r : Y \longrightarrow X \; [ri = 1_X]$$

Such a mapping r is also sometimes called a "splitting for i" instead of "retraction for i" or "left-inverse for i". Here is a picture with internal diagrams (since this is a monomapping, it is already a part of Y):

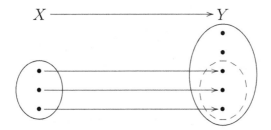

Now for r to be a retraction from Y back to X along i, $ri = 1_X$, we see that for any element x

$$r(ix) = (ri)x = x$$

so any element y of Y that came from an x must be sent by r back to where it came from, and that is *all* the equation $ri = 1_X$ says. It says nothing about the other elements of Y, if there are any; but of course, to be a mapping $Y \longrightarrow X$, r must be *defined* for all those other elements of Y as well, and r must send them somewhere in X (it does not matter where we send them, and so there will usually be many possibilities for the mapping r).

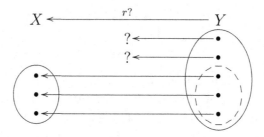

One possibility is

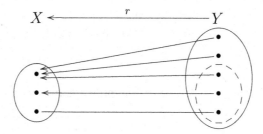

For example, with S the set of students, C the set of chairs, and

$$S \xrightarrow{i} C$$

assigning to each student the chair in which that student is sitting, r must assign a student to *every* chair, and the equation says that to each *occupied* chair must be assigned the student who is occupying it. But for the other chairs we can make any definite decision we want; for instance, r could assign all the unoccupied chairs to one particular student, or any other rule could be adopted for the unoccupied chairs and would nonetheless complete the definition of yet another retraction r for our given i.

We hope this example will show that, in the category of abstract sets, almost every monomapping has at least one retraction: just pick any element of the domain to be the recipient of all the extra elements of the codomain. The only difficulty arises when the domain is *empty* and the codomain is not: the (unique) map

$$0 \longrightarrow Y$$

is a monomapping but is not split if $Y \neq 0$.

Claim 3.4: *In the category of abstract sets and mappings, if $X \overset{i}{\longrightarrow} Y$ is a monomapping with domain X not empty, then i has a retraction.*

To prove the claim we will use two properties of the category of abstract sets and arbitrary mappings whose precise relation to our axioms will be established later in Section 6.1:

(1) If $X \neq 0$, then X has an element $1 \overset{x_0}{\longrightarrow} X$;
(2) every part $X \overset{i}{\longrightarrow} Y$ has a **complement** $X' \overset{i'}{\longrightarrow} Y$, i.e.

$$X \overset{i}{\longrightarrow} Y \overset{i'}{\longleftarrow} X'$$

is a *sum* diagram.

Now, we proceed with claim 3.4; X is not empty, so let $1 \overset{x_0}{\longrightarrow} X$ be an element and $X' \overset{i'}{\longrightarrow} Y$ a complement of i. The defining property of sum says that to define any mapping $Y \longrightarrow X$ (in particular the r we are looking for) is equivalent to specifying mappings from the two parts whose sum is Y. We use the mappings we have,

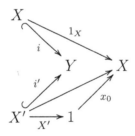

where the bottom arrow is the only mapping $X' \longrightarrow 1$. Now the universal mapping property of sum tells us that there will be a mapping $Y \overset{r}{\longrightarrow} X$, making the diagram commute as follows:

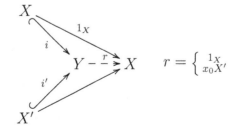

Of course, we do not really need the commutativity of the bottom triangle, for the top triangle says $ri = 1_X$, which is all we wanted.   ■

A proposition of this type is definitely not true in other categories with more geometrical content. Examples of such categories are in the next exercise. We discuss additional examples below in Exercise 8.11.

Exercise 3.5

The category C has as objects those parts of $\mathbb{R}^n$, for various n, which are open. The arrows of C are mappings between the objects that are continuous. (For definitions of open and continuous see any advanced calculus text.) Show that C is a category, that is, define composition and identities and check the axioms. A subcategory D of C is defined when we restrict to mappings that have a derivative. What rule of elementary calculus shows that D has composition? A subcategory of D is obtained by restricting to the *smooth* mappings – those that have all derivatives.   ◊

In the category C, even though the sphere is a retract of punctured space (the open part of $\mathbb{R}^n$ obtained by removing a single point), it is not a retract of the whole (unpunctured) space. It is important to look at examples like this to get a better idea about what part of our reasoning can be used in these more general situations without change (for instance $ri = 1_X$ *always* implies i is a monomapping) and what part is special to the category of abstract sets S (for instance i is a monomapping with nonempty domain implies $\exists r[ri = 1]$). Let us describe the example more precisely:

In n-dimensional space (e.g., $n = 2$, the plane) we have the $(n - 1)$-dimensional sphere (i.e., the circle, if $n = 2$).

Claim 3.6: *There does not exist any continuous retraction for the inclusion:* $(n - 1)$-dimensional unit sphere $\xrightarrow{\;\;i\;\;}$ n-dimensional space, centered at the origin.

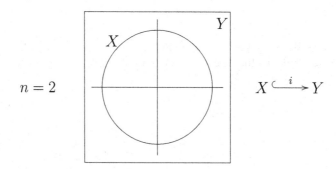

This claim is proved in analysis but is reasonable on the basis of the intuition of continuity. On the other hand, if we change the situation only slightly and replace the codomain Y by the "punctured plane" with the origin removed, then there is a fairly obvious continuous retraction: just send each point in the punctured plane to the unique point on the circle that lies on the same ray from the origin, i.e.,

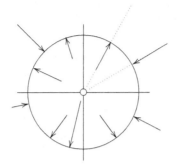

If our circle has radius 1, the formula for r is

$$r(y_1, y_2) = (\frac{y_1}{|y|}, \frac{y_2}{|y|})$$

where $|y| \overset{\text{def}}{=} \sqrt{y_1^2 + y_2^2}$ is the distance from the point $y = (y_1, y_2)$ in the plane to the origin.

We could also have considered as codomain a unit ball rather than the whole Euclidean space with the same result: that the unpunctured version is completely different from the punctured version with respect to the question of continuously retracting onto the sphere.

Exercise 3.7
Suppose $S \overset{i}{\longrightarrow} B \overset{j}{\longrightarrow} E$ are maps in any category. Then i has a retraction if ji does. ◇

That our proof of the ubiquity of retractions for abstract sets does not work for continuous sets can be understood more exactly if we recall that it was based on the concept of sum and still more exactly on the concept of complementary part. Although sums exist generally in the continuous category, it is emphatically not the case that to every part $X_0 \longleftrightarrow Y$ there is another part $X_1 \longleftrightarrow Y$ together with which it sums to Y. Indeed, we might take X_1 as the "rest of" Y after removing X_0, but the defining property of sum, (that any pair of maps f_0, f_1

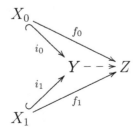

can be combined via definition by cases to give a single map $Y \longrightarrow Z$) will fail in the continuous category $\mathcal{C}$: even if f_0, f_1 were both continuous, their combination on Y would not be unless a further condition were verified. For example, if X_0 is the circle included by i_0 as the boundary of a disc Y and if i_1 is the "rest" (i.e., the open disk included by i_1 as everything in Y except the boundary), then for continuity of the "sum" map $\left\{ \begin{smallmatrix} f_0 \\ f_1 \end{smallmatrix} \right.$ one needs not only continuity of f_0, f_1 separately but also that

$$\lim_{n \to \infty} f_1(y_n) = f_0(x)$$

whenever y_n is a sequence in the open interior i_0, which tends to x on the boundary. These matters are discussed more fully in courses in analysis and geometry.

3.2 Isomorphism and Dedekind Finiteness

"Equations" between sets, such as the statement that a set is the sum of two others or that it is the inverse image of a part along a map (and many others that we will meet soon such as the commutative law for products, the laws of exponents, the distributive law, etc.) are not really equations but rather statements that certain sets have the same number of elements. More precisely, the intended statements often say more: a specific *demonstration* of the equinumerosity is stated to be available. Such a demonstration is a given *isomorphism* in the category of abstract sets and mappings.

Definition 3.8: $X \xrightarrow{f} Y$ *is an* **isomorphism** *(or is* **invertible***) if and only if there exists a* $Y \xrightarrow{g} X$ *for which both equations* $gf = 1_X$ *and* $fg = 1_Y$ *are true. Such a* g *is called a* **(two-sided) inverse** *for* f.

We obviously have the implications

$$f \text{ isomorphism}$$
$$A \Downarrow$$
$$f \text{ has a retraction and } f \text{ has a section}$$
$$B \Downarrow$$
$$f \text{ is injective and } f \text{ is surjective}$$
$$Def \Updownarrow$$
$$f \text{ is bijective}$$

But we claim that implication A is reversible in any category and that implication B is reversible in the category of abstract sets and arbitrary maps; thus, in this category the isomorphisms are the same as the bijections (another name for which is **one-to-one correspondences**).

The reversal of implication A above means that, even though retractions ($=$ left inverses) are not unique, and neither are sections ($=$ right inverses), nonetheless if *both* exist for the same f, then they are all equal:

Proposition 3.9: *In any category,*

$$if \quad X \underset{s}{\overset{r}{\underset{f}{\rightleftarrows}}} Y \quad then$$

$$A^* : \quad rf = 1_X \text{ and } fs = 1_Y \Longrightarrow r = s$$
$$C : \quad \exists r[rf = 1_X] \Longrightarrow \forall s_1 s_2[fs_1 = 1_Y \text{ and } fs_2 = 1_Y \Longrightarrow s_1 = s_2]$$

Proof: On the assumption of the hypothesis of A^*,

$$r = r1_Y = r(fs) = (rf)s = 1_X s = s$$

proving the conclusion of A^*. It should be clear that A^* implies the converse of A. Because A^* is known, the proof of C requires us only to notice that C is equivalent to

$$\forall s_1, s_2 \forall r[rf = 1_X \text{ and } fs_1 = 1_Y \text{ and } fs_2 = 1_Y \Longrightarrow s_1 = s_2]$$

which follows from two applications of A^* together with the remark that

$$r = s_1 \text{ and } r = s_2 \Longrightarrow s_1 = s_2$$

∎

Stated in words, C says that any two sections of f are equal provided that f has at least one retraction; the proviso is certainly needed since most maps that have sections have more than one.

The reversibility of implication B above, namely that any bijection has a two-sided inverse, does not hold in some categories (such as that of continuous maps), but it does hold in case the inverse is allowed to be an "arbitrary" map. One way to prove it would be to assume that every surjective f has at least one section s (this is the axiom of choice; see Section 4.3) and then use the following generally valid proposition.

Proposition 3.10: *If $fs = 1_Y$ and if f is a monomorphism, then s is the two-sided inverse of f.*

Proof: We are supposed to prove that $sf = 1_X$, but as in other cases when we can hope to use a cancellation property, we first prove something a little longer,

$$f(sf) = (fs)f = 1_Y f = f = f1_X$$

from which the conclusion follows by left-cancellation of f. ∎

However, that bijection implies isomorphism does not really depend on a powerful principle such as the axiom of choice; the latter is usually only required when the map g to be constructed is not unique and hence must be chosen arbitrarily. But the bijectivity of f says exactly that

$$\forall y \; \exists! x \, [fx = y]$$

the existence of x coming from the surjectivity of f and the uniqueness from the injectivity of f. Hence, by the concept of "arbitrary map" there should be a well-defined map g such that

$$gy = \text{ the } x \text{ for which } [fx = y]$$

for all y; then of course $gfx = x$ for all x and $fgy = y$ for all y so that g is the two-sided inverse of f because 1 is a separator.

Notation: If $gf = 1_X$ and $fg = 1_Y$, we write $f^{-1} = g$.

Proposition 3.11: *If $X \xrightarrow{f} Y$ is an isomorphism, then $Y \xrightarrow{f^{-1}} X$ is also an isomorphism.*

Proof: Writing the two equations $gf = 1_Y$, $fg = 1_X$ in the opposite order, we see that $f = g^{-1}$, i.e. the f^{-1} has f as a two-sided inverse. ∎

Proposition 3.12: *If each of f_1, f_2 is an isomorphism and if $f_2 f_1$ makes sense, then $f_2 f_1$ is also an isomorphism.*

Proof: Let g_1, g_2 be two-sided inverses

The only obvious candidate for the inverse of the composite $f = f_2 f_1$ is the composite *in opposite order* $g = g_1 g_2$ of the inverses. Indeed

$$gf = (g_1 g_2)(f_2 f_1) = g_1(g_2 f_2)f_1 = g_1 1_Y f_1 = g_1 f_1 = 1_X$$
$$fg = (f_2 f_1)(g_1 g_2) = f_2(f_1 g_1)g_2 = f_2 1_Y g_2 = f_2 g_2 = 1_Z$$

In other words,

$$(f_2 f_1)^{-1} = f_1^{-1} f_2^{-1}$$

is the required two-sided inverse. ∎

Notation: $f : X \xrightarrow{\sim} Y$ means that f is an isomorphism. We say X is *isomorphic* with Y, written $X \cong Y$ if there exists at least one f from X to Y, which is an isomorphism, i.e.

$$\exists f \ [f : X \xrightarrow{\sim} Y]$$

As discussed previously this definition is used in every category, but in the case of the category of abstract sets and arbitrary maps it is often referred to as being **equinumerous** or **having the same cardinality** with isomorphic being reserved for other categories in which naturally some further structure (such as group structure or topological structure) will be the "same" between isomorphic X, Y. However, abstract sets should be considered as the case of "zero structure," and because we have learned to treat the number zero on an (almost) equal footing with other numbers, we should simplify terminology and our thinking by treating the category of sets on an (almost) equal footing with other categories such as are met in more structured parts of mathematics. That isomorphism of abstract sets gives us a way to study equinumerosity *without counting* (hence also for *infinite* sets) was exploited systematically by Cantor. (He was anticipated to some extent by Galileo and Steiner.)

Exercise 3.13
[Galileo Galilei, 1638] The set of all nonnegative whole numbers is isomorphic with the set of all square whole numbers because the map $f(x) = x^2$ has (with the stated domain and codomain) a two-sided inverse g. What is the usual symbol for this g? ◇

Of course counting will be *one* of the methods for investigating size (all methods are consequences of the definition). Namely, if X is thought of as a "standard" set (like $\{1, 2, 3, 4, 5\}$), then a *specific* counting process that counts a set Y is of course a specific invertible map $f : X \xrightarrow{\sim} Y$. (Note that the same Y can be counted in different "orders" by the same X, which would involve different maps.) If Z can also be counted by the same X, we can then conclude that Y and Z are (objectively) the same size (which is true independently of our subjectively chosen standard X and of our choices of counting order):

Corollary 3.14:

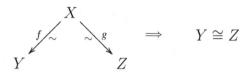

$$\Longrightarrow \qquad Y \cong Z$$

Proof: By Propositions 3.11 and 3.12, $h \overset{\text{def}}{=} gf^{-1}$ shows the required existence. ∎

   The observation of Galileo that an infinite set can be isomorphic to a proper part of itself was turned into a definition by Dedekind:

Definition 3.15: *A set X is said to be* **Dedekind-finite** *if and only if*

$$\forall f[X \overset{f}{\longrightarrow} X \text{ and } f \text{ monic} \implies f \text{ isomorphism}]$$

and correspondingly X is said to be **Dedekind-infinite** *if and only if*

$$\exists f[X \overset{f}{\longrightarrow} X \text{ and } f \text{ monic and } f \text{ not surjective}]$$

Exercise 3.16
Show that $X = [0, 1]$, the unit interval of calculus, is Dedekind-infinite.   ◊

Exercise 3.17
If X is Dedekind-finite, $A \overset{i}{\longhookrightarrow} X$ is a part of X and A is equinumerous with X (i.e., $A \cong X$), then the part is the whole in the sense that i itself is invertible.   ◊

3.3 Cartesian Products and Graphs

There is a common theme involving several different interrelated constructions that we will now explore:

FINITE INVERSE LIMITS

Cartesian operations
$\begin{cases}
\text{Terminal object} \\
\text{Inverse image (of a part along a map)} \\
\text{Intersection (of parts)} \\
\text{Equalizers} \\
\text{Cartesian products} \\
\text{Pullbacks} (= \text{fibered products}) \\
\\
\text{and numerous subsidiary notions like} \\
\text{Graphs, diagonal, projections}
\end{cases}$

The French philosopher René Descartes is said to have introduced the idea that dealing with more than one quantity can be done by introducing higher-dimensional space. Along with that goes the idea of describing parts of higher-dimensional space by means of equations. For example, consider the portion (circle) of the plane where a certain equation is true. The operation of passing from the coordinatized plane to the portion where the equation $x^2 + y^2 = 1$ is true will be an example of the *equalizer* process. We can also intersect figures in higher-dimensional space, raising questions such as: If two figures are described by equations, then is their intersection also described by equation(s)? and so on. There is a rich interplay of all these Cartesian operations, which are also called finite inverse limits.

We will find that *Cartesian products* and *equalizers* will be sufficient to construct inverse limits, and so it turns out that inverse images can be defined in terms of equalizers and Cartesian product. Alternatively, *equalizers* can be defined in terms of inverse images, and thus we will have to understand not only what each of these operations is but also how some of them can be defined in terms of others because these relationships are part of the lore of how the operations are used.

Let us consider first a very simple but important example, namely the product 2×2, where 2 is a two-element set whose elements we can call 0,1 for convenience. It will then turn out that the product 2×2 has four elements, which can be labeled

$$\langle 0, 1 \rangle \qquad \langle 1, 1 \rangle$$
$$\langle 0, 0 \rangle \qquad \langle 1, 0 \rangle$$

by "ordered pairs" of elements of 2. The universal mapping property, which we will require of products in general, amounts in this case to the following: Suppose X is any set and suppose that

$$X \xrightarrow{\varphi} 2, \quad X \xrightarrow{\psi} 2$$

are any two given maps with the same domain X and whose codomains are the factors of the product 2×2 under consideration; then, there is a uniquely defined single map

$$X \xrightarrow{\langle \varphi, \psi \rangle} 2 \times 2$$

whose domain is still X but whose codomain is the product and whose value at any element x is the element of 2×2 labeled by the ordered pair of values determined by the given φ, ψ:

$$\langle \varphi, \psi \rangle x = \langle \varphi x, \psi x \rangle$$

The mapping $\langle \varphi, \psi \rangle$ can be understood in terms of the parts of X for which φ, ψ are the characteristic functions with 1 interpreted as true and 0 as false. If x belongs to the part characterized by φ but does not belong to the part characterized by ψ, then, at such an x, $\langle \varphi, \psi \rangle$ has the value

$$\langle \varphi, \psi \rangle x = \langle 1, 0 \rangle$$

and similarly for the other three kinds of element x definable in terms of φ, ψ. Now the mappings

$$2 \times 2 \longrightarrow 2$$

are usually called propositional operations or truth tables. For example, the map & described by the table $2 \times 2 \xrightarrow{\&} 2$

u	v	u & v
0	0	0
0	1	0
1	0	0
1	1	1

is usually called "conjunction" and read "and". Now a composite

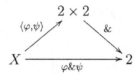

with a propositional operator such as & yields a new property $\varphi \& \psi$ of (elements of) X constructed out of the two given properties φ, ψ. A propositional operation corresponds, via characteristic functions, to an operation on parts of any given X.

Proposition 3.18: *If φ is the characteristic function of the part i of X and ψ is the characteristic function of the part j of X, then*

$$\varphi \ \& \ \psi \ \text{is the characteristic function of} \ i \cap j$$

or, equivalently, if we denote by $\{X \mid \varphi\}$ the part of X with characteristic function φ, we have

$$\{X \mid \varphi \ \& \ \psi\} \equiv \{X \mid \varphi\} \cap \{X \mid \psi\}$$

Exercise 3.19
Prove the proposition.   ◊

   The product of any two sets A, B should be thought of as a "bag of dots" P that has "the right number of elements" and is moreover equipped with a pair of "projection" maps π_A, π_B that "arrange P in rectangular fashion" as follows:

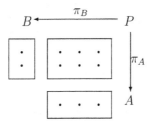

These features will follow from the definition below for products that directly express the most basic act we constantly do in actual use of products. In the category of abstract sets and mappings we will use the same definition of product that we use in any category.

Definition 3.20: *In any category, a* **product** *of two objects A and B is a pair of given mappings*

$$A \xleftarrow{\pi_A} P \xrightarrow{\pi_B} B$$

such that

$$\forall X, f : X \longrightarrow A, g : X \longrightarrow B \; \exists! h : X \longrightarrow P \; [f = \pi_A h \text{ and } g = \pi_B h]$$

as in the following commutative diagram:

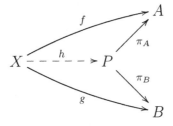

AXIOM: BINARY PRODUCTS
Any two sets A, B have a product, $A \xleftarrow{\pi_A} P \xrightarrow{\pi_B} B$.

Usually we write $P = A \times B$, $h = \langle f, g \rangle$ so that the diagram in the axiom is

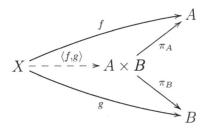

Exercise 3.21

Show directly from the axiom that there is a map

$$B \times A \xrightarrow{\tau_{A,B}} A \times B$$

for which

$$\pi_A^{A,B}\tau_{A,B} = \pi_A^{B,A} \qquad \pi_B^{A,B}\tau_{A,B} = \pi_B^{B,A}$$

Using the uniqueness of the induced map as required in the axiom, show that

$$\tau_{A,B}\tau_{B,A} = 1_{A \times B}$$

the identity mapping. *Caution:* In dealing with more than one product, it is often necessary to use a more precise notation for the projections as we have done in the description of τ.   ◊

Exercise 3.22

Show that for any A, $1 \times A \cong A$.

Hint: To show one of the equations requires using the uniqueness clause in the definition of product.   ◊

The uniqueness of the induced maps into a product enables one to calculate their values at (generalized) elements: If $X \xrightarrow{f} A$, $X \xrightarrow{g} B$, and if $T \xrightarrow{x} X$, consider the diagram

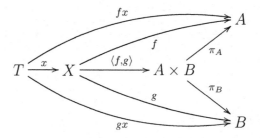

We see that $\langle f, g \rangle x$ has the projections

$$\pi_A(\langle f, g \rangle x) = (\pi_A\langle f, g \rangle)x = fx$$
$$\pi_B(\langle f, g \rangle x) = (\pi_B\langle f, g \rangle)x = gx$$

Hence, by uniqueness of maps with specified projections, we obtain

$$\langle f, g \rangle x = \langle fx, gx \rangle$$

An example apparently less abstract than that of 2×2 can be obtained by considering the description f of the flight of a fly trapped in a tin can C (during a time interval T). If A denotes the points on the base of the can, and if B denotes the points on the side seam, then there is a map $C \xrightarrow{\pi_A} A$ that assigns to each point of the space in the can its "shadow" in A, and there is also $C \xrightarrow{\pi_B} B$ assigning to each point in C the point on the seam at the same height. Then by composition we obtain

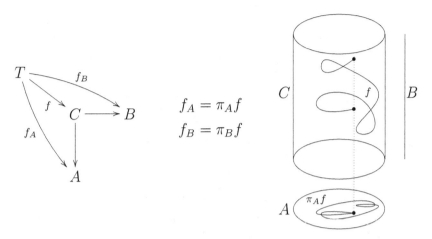

$$f_A = \pi_A f$$
$$f_B = \pi_B f$$

But conversely, if we were given a mapping $T \xrightarrow{f_A} A$ from times to A (for example by a movie film that had been shot with the camera looking up from below through a glass bottom), and if we were similarly given a temporal record $T \xrightarrow{f_B} B$ of the heights, we could uniquely reconstruct the actual flight $T \xrightarrow{f} C$ as the only mapping satisfying the two composition equations.

Definition 3.23: *If $A \xrightarrow{f} B$ is any mapping, then (relative to a given object configured as a product of A and B) by the* **graph** *of f is meant the mapping $A \xrightarrow{\gamma_f} A \times B$ whose projections are the identity map 1_A and f respectively as in the commutative diagram:*

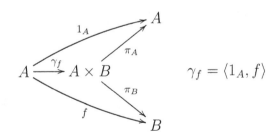

$$\gamma_f = \langle 1_A, f \rangle$$

For example, for any element x of A the value of γ_f is $\gamma_f(x) = \langle x, fx \rangle$ in the product:

Proposition 3.24: *The graph of f is a part of $A \times B$. The elements of $A \times B$, which are members of the part γ_f, are precisely all those whose form $\langle x, y \rangle$ has the property that $y = fx$.*

Proof: Since $\pi_A \gamma_f = 1_A$, we have that π_A is a retraction for γ_f and hence γ_f is monic and therefore a part of its codomain $A \times B$. If $T \xrightarrow{p} A \times B$ is any element of $A \times B$, it is necessarily of the form

$$p = \langle x, y \rangle$$

where $x = \pi_A p$, $y = \pi_B p$. If $y = fx$, then $\gamma_f(x) = \langle x, fx \rangle = \langle x, y \rangle = p$ so that

$p \in \gamma_f$

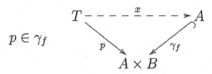

Conversely, if p is a member of γ_f there must be an x that "proves" this membership so that $p = \langle x, fx \rangle$. ∎

Using the picture of a product as a rectangular arrangement with pairs labeling its elements, we get a very graphical internal picture of a map that is at least as important as the cographical internal picture of f.

$\gamma_f x = \langle x, fx \rangle$

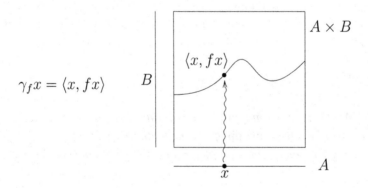

When we need to calculate explicitly using this idea that the graph is *also* a map, we recover the original f just by composing

$$f = \pi_B \gamma_f$$

Proposition 3.25: *Any section of the projection $A \times B \xrightarrow{\pi_A} A$ is the graph of a unique map $A \longrightarrow B$.*

Proof: Recall that a section of a mapping π is a mapping s for which $\pi s = 1_A$. If $\pi = \pi_A$ is the projection from a product we can compose a given section s with the *other* projection, obtaining

$$f \stackrel{\text{def}}{=} \pi_B s$$

But then s must be the graph of the f so defined because

$$\gamma_f = \langle 1_A, f \rangle$$

is the unique map for which the projections are 1_A and f, and by construction s has those two projections. ∎

Proposition 3.25 suggests considering that for any map

$$E \xrightarrow{\pi} A$$

(not necessarily a projection from a product), any section s of π could be regarded as a "generalized map" with domain A but with a *variable* codomain, that is, the value of such a "map" at $1 \xrightarrow{x} A$ lies in the fiber E_x of π over x, and these fibers may vary. This point of view is very important in mathematics when, for example, considering vector functions, fiber bundles, and so forth.

Exercise 3.26
Show that the fiber (= inverse image of x along π) of π at x "is" B if $\pi = \pi_A : A \times B \longrightarrow A$. ◇

In the case of a product with two (or more) factors repeated, we may use a slightly different way of abbreviating the notation for the projections:

$$A \times A = A^2 \underset{\pi_1}{\overset{\pi_0}{\rightrightarrows}} A \qquad \pi_0 \langle a_0, a_1 \rangle = a_0 \qquad \pi_1 \langle a_0, a_1 \rangle = a_1$$

Moreover, among the many possible graphs in this case, there is an especially simple one that nonetheless sometimes needs to be mentioned explicitly, the graph of the identity map on A, which is often called the **diagonal** map $\delta_A = \langle 1_A, 1_A \rangle$, so

$$A \xrightarrow{\delta_A} A \times A \underset{\pi_1}{\overset{\pi_0}{\rightrightarrows}} A \quad \text{satisfies} \quad \pi_k \delta_A = 1_A \text{ for } k = 0, 1$$

A given map

$$A^2 \xrightarrow{\theta} A$$

is often called a "binary algebraic operation on A" since, for any two elements a_0, a_1 of A, θ will produce a single element as a result:

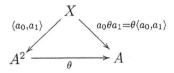

(A frequently used notation is to write the symbol for the operation between the symbols for the two elements of A, which are the components of the element of A^2 to which the operation is being applied). Thus, for example & is a binary operation on $A = 2$ (the set of truth values), and multiplication $\mathbb{R}^2 \longrightarrow \mathbb{R}$ is a binary operation on the set $A = \mathbb{R}$ of real numbers.

Exercise 3.27
What is the composite map

$$\mathbb{R} \xrightarrow{\;\delta_R\;} \mathbb{R}^2 \overset{\cdot}{\underset{?}{\longrightarrow}} \mathbb{R}$$

usually called? ◊

Exercise 3.28
A binary operation θ is said to satisfy the "idempotent law" if the composite

$$A \xrightarrow{\;\delta_A\;} A^2 \overset{\theta}{\underset{1_A}{\longrightarrow}} A$$

is the identity. Does multiplication of real numbers satisfy the idempotent law? What about conjunction $2 \times 2 \longrightarrow 2$ of truth values? ◊

3.4 Equalizers

Definition 3.29: *Given* $A \underset{g}{\overset{f}{\rightrightarrows}} B$, *an* **equalizer** *for f with g is any map* $E \xrightarrow{\;i\;} A$ *such that*

(1) $fi = gi$
(2) $\forall T \, \forall x \, [fx = gx \implies \exists! \, \bar{x} \, [x = i\bar{x}]]$

as in the following commutative diagram:

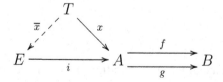

Exercise 3.30
Any equalizer i (as in the definition) is a part of A. If i is an equalizer, then for any x

$$x \in i \iff fx = gx$$

◊

Exercise 3.31

Suppose that $E \xrightarrow{i} A$ and $E' \xrightarrow{i'} A$ are equalizers of the *parallel pair* $A \underset{g}{\overset{f}{\rightrightarrows}} B$. Show that $E \cong E'$ by an isomorphism compatible with i, i'; we say that i and i' themselves are isomorphic.

Hint: Use the universal mapping property of equalizers to find candidate inverse isomorphisms. ◊

Notice that the *same* proof would also show that any two products of X and Y are isomorphic. This justifies a common abuse of language that allows us to speak of *the* equalizer of f and g or *the* product of X and Y. That is, any two equalizers (or products) are "equal" *up to isomorphism*, and this provides further examples of the "equations" among sets that we mentioned at the beginning of Section 3.2. The proof that the inverse image of a part along a map is unique up to isomorphism is the same as those given above. The proof that the sum is unique up to isomorphism is *dual*, i.e., it is the same proof with the arrows reversed.

Because of their great importance in mathematics, we might want to assume as an axiom that equalizers always exist for any parallel pair f, g of maps. However, having assumed the existence of products and inverse images, it turns out that we can prove the existence of equalizers – indeed by two different constructions.

Exercise 3.32

The equalizer of the projection maps $B^2 \underset{\pi_1}{\overset{\pi_0}{\rightrightarrows}} B$ is the diagonal $B \xrightarrow{\delta_B} B^2$. Given any $A \underset{g}{\overset{f}{\rightrightarrows}} B$, we can form the single map $\langle f, g \rangle : A \longrightarrow B^2$ along which we can take the inverse image of the part δ_B.

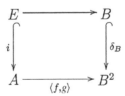

Prove that if i is assumed to have the universal mapping property appropriate to the particular inverse image, it follows that i also has the universal mapping property required of the equalizer of f with g. ◊

The other construction of the equalizer will only use that special case of inverse image known as intersection. Given $A \underset{g}{\overset{f}{\rightrightarrows}} B$, then each of them has a *graph*, which is a part of $A \times B$:

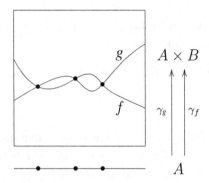

If we *intersect* these two parts of $A \times B$, we clearly get a part of $A \times B$ that "sits over" the desired part of A. To establish this fact rigorously, note that both of the graphs actually have the same domain A, thus, their intersection $E \xrightarrow{j} A \times B$, being included ($\subseteq$) in both, participates in a commutative diagram

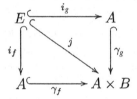

where i_f and i_g are the "proofs" of the respective inclusions. But i_f, i_g are both maps from E to A, and the unusually special construction actually leads to the following:

Proposition 3.33: $i_f = i_g$.

Proof: Since $\gamma_f i_f = \gamma_g i_g$,

$$\pi_A \gamma_f i_f = \pi_A \gamma_g i_g$$

But by half of the definition of graph,

$$\pi_A \gamma_f = 1_A = \pi_A \gamma_g$$

hence, $i_f = i_g$.   ∎

Exercise 3.34
Define $i = i_f = i_g$ as in Proposition 3.33. Show that $fi = gi$ and that indeed i is the equalizer of f with g. Hence, the equalizer of two maps can always be constructed whenever we can form their graphs and intersect them.   ◊

3.5 Pullbacks

The constructions of terminal object, cartesian product, inverse image, intersection, and equalizer are all special cases of one general construction called "limit". In practice there are several other instances that often occur, a frequent one being "pullback" or "fibered product". We will show in Section 3.6 that the existence of equalizers and finite products suffices to guarantee the existence of all finite limits and also that the existence of pullbacks and the terminal object suffices for the existence of all finite limits.

Definition 3.35: *Given any two maps with common codomain* $A_0 \xrightarrow{f_0} B \xleftarrow{f_1} A_1$ *a* **pullback** *(or* **fibered product***) of them means a pair of maps* π_0, π_1 *with common domain P that*

(1) *form a commutative square* $f_0\pi_0 = f_1\pi_1$

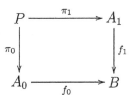

and, moreover,

(2) *are* universal *among such squares in that,*
for any T, a_0, a_1

$$f_0a_0 = f_1a_1 \Rightarrow \exists!a \ [a_0 = \pi_0a \ and \ a_1 = \pi_1a]$$

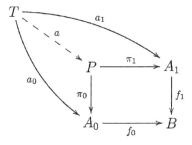

We point out that like products, equalizers, inverse images, and sums, pullbacks are unique up to unique isomorphism. The proof is once again that of Exercise 3.31, and we will speak of P as *the* pullback of f_0 and f_1 without explicitly mentioning π_0 and π_1.

The uniqueness of the induced maps into a pullback enables one to calculate their values at (generalized) elements: Suppose $X \xrightarrow{a_0} A_0, X \xrightarrow{a_1} A_1$ satisfy

$f_0 a_0 = f_1 a_1$, and so they define an arrow to the pullback denoted $X \xrightarrow{\langle a_0, a_1 \rangle} P$. If $T \xrightarrow{x} X$ is any generalized element of X, then the value $\langle a_0, a_1 \rangle x$ has the projections

$$\pi_0(\langle a_0, a_1 \rangle x) = (\pi_0 \langle a_0, a_1 \rangle) x = a_0 x$$
$$\pi_1(\langle a_0, a_1 \rangle x) = (\pi_1 \langle a_0, a_1 \rangle) x = a_1 x$$

Hence, by uniqueness of maps with specified projections, we obtain

$$\langle a_0, a_1 \rangle x = \langle a_0 x, a_1 x \rangle$$

and hence the latter pair is the generalized element of P, which is the required value.

Proposition 3.36: *If products of two objects and equalizers exist, then pullbacks exist.*

Proof: Given data f_0, f_1 as above, first form the product $A_0 \times A_1$ with projections now denoted p_i for $i = 0, 1$. The square

$$
\begin{array}{ccc}
A_0 \times A_1 & \xrightarrow{\ \ p_1\ \ } & A_1 \\
{\scriptstyle p_0}\downarrow & & \downarrow{\scriptstyle f_1} \\
A_0 & \xrightarrow[\ \ f_0\ \ ]{} & B
\end{array}
$$

is usually *not* commutative, but we can form the equalizer $P \xhookrightarrow{\ i\ } A_0 \times A_1$ of the two composites $f_0 p_0$, $f_1 p_1$, that is the *part* of the product on which the diagram *becomes* commutative. If we then define

$$\pi_0 = p_0 i \qquad \pi_1 = p_1 i$$

we will have $f_0 \pi_0 = f_1 \pi_1$. If the (generalized) element $\langle a_0, a_1 \rangle$ of $A_0 \times A_1$ satisfies $f_0 a_0 = f_1 a_1$, then it will actually be a member of i, with a unique "proof" a; but

$$\langle a_0, a_1 \rangle = ia \Rightarrow$$
$$a_0 = p_0 \langle a_0, a_1 \rangle = p_0 ia = \pi_0 a$$
$$a_1 = p_1 \langle a_0, a_1 \rangle = p_1 ia = \pi_1 a$$

as required of a pullback.   ■

Proposition 3.37: *The existence of pullbacks and terminal object implies the existence of binary products, inverse images, intersections, and equalizers.*

Proof: If $B = 1$, then there is only one possibility for f_0, f_1, and moreover any square ending in B is commutative. Hence, the diagram

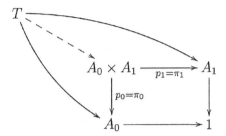

shows that product is literally a special case of pullback provided we have a terminal object 1 (of course in the category of sets we have a terminal object, namely a one-element set). We have previously seen that equalizers can be constructed using products and intersections (of graphs) *or* using products and inverse images (of diagonal parts). Further, we saw that intersection may be considered as a special case of inverse image. The proposition will be proved if we can show that inverse image is a special case of pullback; the interesting aspect, which requires proof, is considered in the following exercise.

Exercise 3.38
Suppose that

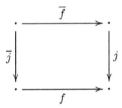

is a pullback square and also that j is a monomorphism. Then it follows that $\overline{j}$ is a monomorphism. ∎

3.6 Inverse Limits

Now we want to describe the more general concept of limit. The data for a limit construction involve a family A_i of objects and a family f_j of arrows between them. Actually, we should specify a **data type**

$$J \mathrel{\substack{\xrightarrow{d} \\ \xrightarrow[c]{}}} I$$

(or "directed graph"), where J is a set of indices for the arrows f_j in specific data of this type and I is the set of indices for the objects A_j. In data of the given type, the domains and codomains are required to be given by the maps d, c:

$$A_{d(j)} \xrightarrow{f_j} A_{c(j)} \quad \text{for all} \quad 1 \xrightarrow{j} J$$

(We call the data type *finite* if I and J are finite. We do not require that the A_i's be finite.)

Definition 3.39: *The* **limit** *(also called* **inverse limit***) of given data A_i, f_j as above is given by a single object L together with a* **universal cone***; here by a* **cone** *with vertex L and with base the given data is meant a family*

$$L \xrightarrow{\pi_i} A_i \quad \text{for} \quad 1 \xrightarrow{i} I$$

of maps satisfying

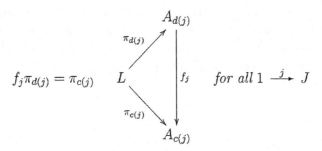

$$f_j \pi_{d(j)} = \pi_{c(j)} \qquad \text{for all } 1 \xrightarrow{j} J$$

That the above cone is **universal** *means that for any object T and any cone with vertex T and with the given data as base, i.e. for any family*

$$T \xrightarrow{a_i} A_i \quad \text{for} \quad 1 \xrightarrow{i} I$$

for which

$$f_j a_{d(j)} = a_{c(j)} \quad \text{for all} \quad 1 \xrightarrow{j} J$$

there is a unique

$$T \xrightarrow{a} L$$

that preserves the cone structure such that

$$a_i = \pi_i a \quad \text{for all} \quad 1 \xrightarrow{i} I$$

For example, products of I factors are easily seen to be limits according to the data type

$$0 \rightrightarrows I$$

for in the case of products there are no given "bonding maps" f_j and correspondingly no equations presupposed in the universal property. On the other hand, equalizers are limits according to the data type

$$2 \underset{c}{\overset{d}{\rightrightarrows}} 2$$

where d, c are the two distinct *constant* maps, because any equalizer data

$$A \underset{g}{\overset{f}{\rightrightarrows}} B$$

involves two given arrows between two given objects according to the scheme that both arrows have the *same* domain A (hence constancy of d) but also the *same* codomain B (hence constancy of c).

Exercise 3.40
Specify a pair of mappings $2 \rightrightarrows 3$ so that limits according to this data type are pullbacks. $\diamond$

Once again we note that *limits are unique up to isomorphism.*

AXIOM: FINITE INVERSE LIMITS
$\mathcal{S}$ has all finite limits.

This means that there is a limit for any data for any finite data type whatsoever. Luckily the general property can be deduced from simpler information.

Theorem 3.41: *In any category, if products and equalizers exist, then limits according to any data type can be constructed.*

To help in understanding the following proof notice that Proposition 3.36 is a special case of this theorem and that the construction used in the proof of that proposition is generalized here.

Proof: Given the data type

$$J \underset{c}{\overset{d}{\rightrightarrows}} I$$

and also given A_i, f_j such that

$$A_i \xrightarrow{f_j} A_{i'} \quad \text{whenever} \quad \begin{array}{l} d(j) = i \\ c(j) = i' \end{array}$$

consider the products $\prod\limits_{1\xrightarrow{i}I} A_i$ and also $\prod\limits_{1\xrightarrow{j}J} A_{c(j)}$.

Thus, for example, if the data type is

$$4 \Longrightarrow 5$$

expressing the graph

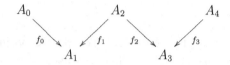

then the first product has five factors, namely all the A_i's, but the second product has four factors that (since c has only two values, etc.) are really just A_1 taken twice and A_3 taken twice.

Now between the two products we can construct two maps:

$$\prod_{1\xrightarrow{i}I} A_i \overset{p}{\underset{f}{\rightrightarrows}} \prod_{1\xrightarrow{j}J} A_{c(j)}$$

where for all j

$$
\begin{array}{ccc}
\Pi A_i & \xrightarrow{\ f\ } & \Pi A_{c(j)} \\
{\scriptstyle p_{d(j)}}\downarrow & & \downarrow{\scriptstyle p_j} \\
A_{d(j)} & \xrightarrow{\ f_j\ } & A_{c(j)}
\end{array}
\qquad p_j f = f_j p_{d(j)}
$$

which determines f by determining all of its J coordinates as being (essentially) all the given bonding maps f_j. The other map p is simpler for it does not depend on the given bonding maps at all.

$$
\begin{array}{ccc}
\Pi A_i & \xrightarrow{\ p\ } & \Pi A_{c(j)} \\
{\scriptstyle p_{c(j)}}\downarrow & & \downarrow{\scriptstyle p_j} \\
A_{c(j)} & \xrightarrow{\ 1_{A_{c(j)}}\ } & A_{c(j)}
\end{array}
\qquad p_j p = p_{c(j)}
$$

Now if we are given any family of maps

$$T \xrightarrow{\ a_i\ } A_i \quad \text{for} \quad 1 \xrightarrow{\ i\ } I$$

it can be considered as a single map $T \xrightarrow{\langle a_i \rangle_I} \Pi_I A_i$. Consider the diagram

$$T \xrightarrow{\langle a_i \rangle_I} \Pi_I A_i \underset{f}{\overset{p}{\rightrightarrows}} \Pi_J A_{c(j)}$$

By construction of p and f, $\langle a_i \rangle_I$ will satisfy the equation

$$p \langle a_i \rangle_I = f \langle a_i \rangle_I$$

if and only if the given family satisfies the family of compatibility equations

$$a_{i'} = f_j a_i \quad \text{whenever} \quad \begin{matrix} d(j) = i \\ c(j) = i' \end{matrix}$$

Hence, if we let

$$L \xrightarrow{e} \Pi A_i \qquad \text{be an equalizer of } p, f$$

then precisely such "f_j compatible" families will factor across the monic e. That is, if we define

$$\pi_i = p_i e \quad \text{for} \quad 1 \xrightarrow{i} I$$

then π_i will have the universal mapping property required of the limit. ∎

Corollary 3.42: *If pullbacks and a terminal object exist in a category, then all finite limits exist.*

Proof: Proposition 3.37. ∎

Exercise 3.43
A terminal object is an inverse limit.

Hint: The data type for the limit has both I and J empty. ◊

Exercise 3.44
For any given single map $X \xrightarrow{f} Y$, the pullback of f along itself is often called "the equivalence relation associated to f". This pullback involves a pair of maps $K \rightrightarrows X$ and hence a single map e to $X \times X$. For any parallel pair $T \overset{x_1}{\underset{x_2}{\rightrightarrows}} X$, $f x_1 = f x_2$ iff $\langle x_1, x_2 \rangle$ belongs to e. (See also Proposition 4.8 and Exercises 4.22 and 4.23.) ◊

3.7 Additional Exercises

Exercise 3.45
Show that the category of finite-dimensional vector spaces has all finite limits.

Hint: They are computed using the finite limits in $\mathcal{S}$.

Exercise 3.46

(a) Show that in the category of finite-dimensional vector spaces the sum and the product of two vector spaces V, V' are the *same* vector space that we usually denote $V \oplus V'$. It has both projections *and* injections.

This means that if V is a vector space there is both a diagonal linear transformation $V \xrightarrow{\delta_V} V \oplus V$ and a *codiagonal* linear transformation $V \oplus V \xrightarrow{\gamma_V} V$. Therefore we can define a binary operation on linear transformations $T, T' : V \longrightarrow V'$, whose result is the threefold composite

$$T + T' = V \xrightarrow{\delta_V} V \oplus V \longrightarrow V' \oplus V' \xrightarrow{\gamma_{V'}} V'$$

The middle arrow is called $T \oplus T'$.

(b) Show how to define $T \oplus T'$, that $+$ coincides with the sum of linear transformations met in linear algebra, and that $+$ makes the linear transformations from V to V' into a group.

Exercise 3.47

Show that the category of groups has all finite limits.

Exercise 3.48

(a) Show that the categories $\mathcal{S}/X$ have all finite limits.
   Hint: They are computed "fiberwise".

More generally,

(b) for any category **C** that has all finite limits, show that **C**$/X$ has all finite limits for any X in **C**.

Exercise 3.49

Show that the category of pointed sets $1/\mathcal{S}$ has all finite limits.

Exercise 3.50

Sets with action: Given a set A (frequently referred to as an "alphabet"), by an "action" of A on a set S is meant any mapping $S \times A \xrightarrow{\delta} S$.

(a) Show that sets equipped with given A-actions form a category if we take as arrows from (S, δ) to (S', δ') those mappings $S \xrightarrow{\varphi} S'$ such that $\varphi\delta = \delta'(\varphi \times 1_A)$. (A special case where A has two elements was considered in Exercise 1.30 (d).)
   When δ is given, it is convenient to abbreviate $\delta(s, a)$ simply by sa and call it the **action** of a on s.

(b) Show that the category of A-sets has sums and finite limits.

Exercise 3.51

M-sets: A **monoid** is a set M with an associative binary operation (that is $m(m'm'') = (mm')m''$ for any three elements) and an element 1 which is an identity (that is $1m = m = m1$ for any m). If M is a monoid and X is a set, a *(right) action* of M on X is a mapping $X \times M \xrightarrow{\delta} X$ such that $x1 = x$ and $(xm)m' = x(mm')$ (where we are writing xm for $\delta(x, m)$). A set X equipped with an action of M is called an *M-set.* A morphism of M-sets from X to X' is a mapping $X \xrightarrow{\varphi} X'$ that is **equivariant**, (i.e., it satisfies an equation like that above for A-sets.)

(a) Show that M-sets form a category.

(b) Show that the category of M-sets has sums and finite limits.

Actually the category of A-sets is a special example of a category of M-sets, where M is the special **free monoid**

$$M = A^* = 1 + A + A \times A + A \times A \times A + \ldots$$

What should the infinite "sum" mean?

Exercise 3.52

If **X** is the category formed from a partially ordered set X, show that a product of two objects is exactly the same thing as a greatest lower bound.

4

Colimits, Epimorphisms, and the Axiom of Choice

4.1 Colimits are Dual to Limits

Each notion of limit has a corresponding "dual" notion of **colimit** whose universal mapping property is obtained by reversing all arrows. Thus, for example, the *terminal* object 1 has the limit property that every object X maps uniquely *to* 1, whereas the *coterminal* (or *initial*) object 0 has the dual property that every object Y has a unique map *from* 0. The limit concept of product, i.e. a **span** (= pair of arrows with common domain) $A \xleftarrow{\pi_A} A \times B \xrightarrow{\pi_B} B$ such that spans $A \xleftarrow{f} X \xrightarrow{g} B$ as below uniquely determine the $\langle f, g \rangle$ in

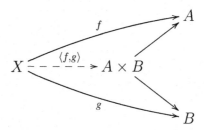

(which expresses the universal mapping property of projections), dualizes to the concept of **coproduct**, i.e. a *cospan* of *injections* $A \xrightarrow{i_A} A + B \xleftarrow{i_B} B$ such that cospans $A \xrightarrow{f} Y \xleftarrow{g} B$ as below uniquely determine the $\left\{ \begin{smallmatrix} f \\ g \end{smallmatrix} \right.$ in

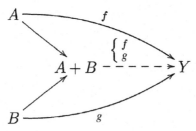

The universal mapping property for injections to the coproduct should be recognized immediately as that for the injections to the *sum*. Similarly, the concept of *equalizer*

for f, g, namely a (necessarily mono!) arrow $E \xrightarrow{i} A$ for which arrows $X \xrightarrow{x} A$ "equalizing" f and g ($fx = gx$) uniquely determine the $X \xrightarrow{\bar{x}} E$, as in

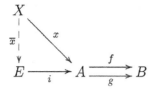

dualizes to the concept of **coequalizer** for f, g, namely an arrow $B \xrightarrow{q} Q$ for which arrows $B \xrightarrow{y} Y$ "*co*equalizing" f and g (i.e., $yf = yg$) uniquely determine the $Q \xrightarrow{\bar{y}} Y$ in

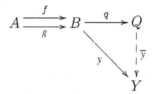

which we will study in more detail later.

Exercise 4.1
Formulate the notion of "pushout," the colimit concept dual to pullback. ◇

Exercise 4.2
Formulate the general notion of the colimit of a data type

$$J \rightrightarrows I$$

◇

Exercise 4.3
Prove that the existence of pushouts and an initial object implies the existence of coproducts.

Hint: Use the dual of Proposition 3.37. ◇

Exercise 4.4
Find sufficient conditions to guarantee the existence of all finite colimits. Prove that the conditions are (necessary and) sufficient.

Hint: Use the dual of Proposition 3.41. ◇

We require the axiom that the category of abstract sets and mappings has all finite colimits:

AXIOM: FINITE COLIMITS IN $\mathcal{S}$

$\mathcal{S}$ has all finite colimits.

This axiom is dual to the axiom of finite limits. It includes the axiom of the empty set and guarantees that finite sums, coequalizers, and pushouts all exist in $\mathcal{S}$.

4.2 Epimorphisms and Split Surjections

Although seemingly not a limit notion, the concept of monomorphism also has its dual, that of epimorphism:

Definition 4.5: *An **epimorphism** f is a map that has the right-cancellation property $\forall \phi, \psi [\phi f = \psi f \Rightarrow \phi = \psi]$.*

We will use the term **epimapping** (or epimap) interchangeably with the term epimorphism for mappings in $\mathcal{S}$ and also use the term **epic** as an adjective.

Exercise 4.6

Any coequalizer is an epimorphism.

Hint: Recall the dual proof! $\Diamond$

Exercise 4.7

Recall the definition of cograph from Section 2.1. Show that the cograph is an epimorphism. Show that any retraction of the injection $B \xrightarrow{i_1} A + B$ is the cograph of a unique mapping from A to B. (These results are dual to Propositions 3.24 and 3.25.) $\Diamond$

We will study below the relationship between epic and "surjective" (which in form is *not* dual to "injective," although it does turn out to be equivalent to the dual in very particular categories such as $\mathcal{S}$).

There is actually an important link between cancellation properties and limits.

Proposition 4.8: *A map i is a monomorphism if and only if when we form the pullback of i with itself the two projections are* equal.

Proof: Suppose the square below is a pullback and moreover $\pi_0 = \pi_1 = \pi$.

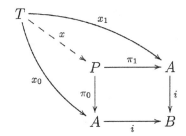

If $i x_0 = i x_1$, then by the universal property of pullbacks, there is a unique $T \xrightarrow{x} P$ for which both $x_0 = \pi_0 x$, $x_1 = \pi_1 x$. But since we have assumed $\pi_0 = \pi_1$, this yields $x_0 = x_1$; i.e. i has left-cancellation.

Conversely, assume i has cancellation and that π_0, π_1 form a pullback of i with i. Then in particular $i\pi_0 = i\pi_1$, hence by cancellation $\pi_0 = \pi_1$. ∎

Exercise 4.9
Without considering the internal picture, show that a map is epimorphic iff two normally different maps in the pushout of the map with itself are equal. ◊

Using simple properties of the sets 1 and 2, one can show that in the category of abstract sets and mappings:

Proposition 4.10: *A mapping $X \xrightarrow{p} Y$ is an epimapping iff it is surjective.*

Proof: Assume p is epic, that is, it satisfies the right-cancellation law

$$\psi_0 p = \psi_1 p \Rightarrow \psi_0 = \psi_1 \qquad X \xrightarrow{p} Y \underset{\psi_2}{\overset{\psi_1}{\rightrightarrows}} V$$

for all V. We want to show that p is surjective, that is

$$\forall y \exists x [px \overset{?}{=} y]$$

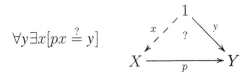

where y is any element of Y with domain 1. To arrive at a proof by contradiction, suppose that contrary to the desired conclusion there is at least one y_0 for which there is no corresponding x, i.e. $px \neq y_0$ for all x. Take $V = 2$ and let ψ_1 be the characteristic function of (the one-element part) y_0, but let ψ_0 be the "constantly false" map $Y \longrightarrow 2$. Then for all elements $x : 1 \longrightarrow X$ of X

$$(\psi_1 p)(x) = \psi_1(px)$$
$$= \text{false since } px \neq y_0 \text{ and } \psi_1(y) = \text{true iff } y = y_0$$
$$= \psi_0(px)$$
$$= (\psi_0 p)x$$

hence, $\psi_1 p = \psi_0 p$ since 1 is a separator. But then $\psi_1 = \psi_0$ since we assumed p epic. However, by the way we defined ψ_0 and ψ_1, $\psi_1 y_0 \neq \psi_0 y_0$, and so we reach a contradiction, showing that there is no such y_0 (i.e., p is surjective).

For the converse part of the proposition, assume that p is surjective and try to show the right-cancellation property; thus, suppose $\psi_1 p = \psi_0 p$, where V is now arbitrary, as are ψ_0, ψ_1. We will use the fact that 1 is a separator. Let y be an arbitrary element of Y. By surjectivity of p there is an element $x : 1 \longrightarrow X$ such that $y = px$, and so $\psi_0 y = \psi_0(px) = (\psi_0 p)x = (\psi_1 p)x = \psi_1(px) = \psi_1 y$.

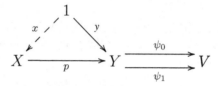

Hence, $\psi_0 = \psi_1$ since 1 separates. Because this cancellation holds for all V, we have proved that p satisfies right-cancellation, i.e. it is epic.   ∎

If we strengthen the notion of surjectivity to demand the extreme case of existence of prevalues for all *generalized* elements, not only elements with domain 1, we obtain the strong condition of *split surjectivity*:

Definition 4.11: *An arrow $X \xrightarrow{p} Y$ is* **split surjective** *if and only if*

$$\forall T \forall y \exists x [px = y]$$

One often expresses this property of p by saying that every y has a **lift** x. To see why, redraw the triangle with p vertical as is often done for continuous mappings in topology.

Proposition 4.12: *An arrow $X \xrightarrow{p} Y$ is split surjective iff there exists a section for p.*

Proof: If p is split surjective, let $T = Y$ and $y = 1_Y$, the identity mapping of Y. Then, as a very special case of the split surjectivity, there exists $x = s$ such that

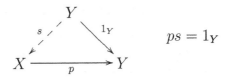

$$ps = 1_Y$$

In other words, p has a section s (usually there will be many).

Conversely, suppose p has a section s; then we are to show that p is split surjective. So consider any T and any $T \xrightarrow{y} Y$. There is an obvious way to attempt to construct the x we need: try $x \stackrel{\text{def}}{=} sy$

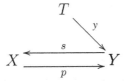

We then have to show that the x so constructed really maps back to y under p, as required for split surjectivity:

$$
\begin{aligned}
px &= p(sy) && \text{by definition of } x \\
&= (ps)y && \text{by associativity} \\
&= 1_Y y && \text{since } s \text{ was a section of } p \\
&= y
\end{aligned}
$$

Thus p is split surjective if it has a section, and hence the proof of the proposition is complete. ∎

Since split surjectivity trivially implies surjectivity (just take the special case $T = 1$), we have, by means of the logical equivalences in the last two propositions, actually already proved the following proposition. However, we will also give a direct proof in the hope of further clarifying the relationship between the concepts. Note that the proof that follows is valid in any category.

Proposition 4.13: *Any mapping p that has a section is an epimorphism.*

Proof: Let s be a section for p. We must show that p has the right-cancellation property; thus, assume

$$\psi_0 p = \psi_1 p \qquad X \xrightarrow{p} Y \underset{\psi_1}{\overset{\psi_2}{\rightrightarrows}} V$$

where V is arbitrary. But then

$$
\begin{aligned}
(\psi_0 p)s &= (\psi_1 p)s \\
\psi_0(ps) &= \psi_1(ps) \\
\psi_0 &= \psi_1 && \text{since } ps = 1_Y
\end{aligned}
$$

as was to be shown. ∎

4.3 The Axiom of Choice

The converse of the last proposition, namely that there are no epimappings except for those that actually have sections s, is not at all obvious. Once we have a section in hand, cancellation becomes the result of a concrete process of calculation, as in the preceding proof, unlike the abstract leap involved in just "doing the cancellation". On the other hand, where the section comes from may also be mysterious. The usual point of view is that the converse is true for constant abstract sets (and is in fact a strong testimony to the "arbitrariness" of the maps in that category), whereas it is obviously *false* for sets that are less abstract, less constant, or both (as we will soon see). The name of the converse is

AXIOM: THE AXIOM OF CHOICE
Every surjective mapping is split surjective (i.e., every epimap has a section).

Much ink has been spilled over this axiom.

Proposition 4.14: *Any section s for a map p is a single procedure that simultaneously chooses an element from each* fiber *of p*

where the important notion of fiber is given by

Definition 4.15: *For each $X \xrightarrow{p} Y$ and $1 \xrightarrow{y} Y$, the* **fiber of** p **over** y *is the domain of inverse image of the singleton (= one element) part y along p*

$$
\begin{array}{ccc}
X_y & \xrightarrow{\ p^{-1}[y]\ } & X \\
\downarrow & & \downarrow{\scriptstyle p} \\
1 & \xrightarrow[\ y\ ]{} & Y
\end{array}
$$

Proof: (of Proposition 4.14.) Since $ps = 1_Y$, $p(sy) = y$, i.e. $sy \in p^{-1}[y]$; in other words sy is a member of the fiber of p over y, but this is so for all y.   ■

Now (as further axioms of higher set theory reinforce) any reasonable family of sets parameterized by a set Y can be realized as the family of fibers of some single map p with codomain Y. The sets in the family are all nonempty if and only if p is surjective (see Exercise 4.16). Hence, the axiom of choice, "Every surjective map is split surjective," says that any family of nonempty sets has at least one choice map that chooses an element from each set in the family.

Exercise 4.16
Show that the mapping p is surjective iff each of its fibers has at least one element.
◊

Exercise 4.17
Show that the mapping p is injective iff each of its fibers has at most one element.
◊

Thus, a mapping is bijective if and only if each of its fibers has exactly one element.

Exercise 4.18
Show that the mapping from real numbers to real numbers defined by the formula

$$f(x) = x^2(x - 1)$$

is surjective and hence has sections in the category of abstract sets and arbitrary mappings. Show, however, that none of these sections are in the continuous category (see Exercise 3.5) because each one must have a "jump".
◊

As we will see in Appendix B, an equivalent form of the axiom of choice, often used in proofs in analysis and algebra, is

The Maximal Principle of Zorn

4.4 Partitions and Equivalence Relations

The dual notion (obtained by reversing the arrows) of "part" is the notion of *partition*.

Definition 4.19: *A **partition** of A is any surjective mapping p with domain A. The fibers of p are called the* **cells** *of the partition.*

Proposition 4.20: *Let* $A \xrightarrow{p} I$ *be surjective, and for each* $1 \xrightarrow{i} I$ *let* A_i *be the cell of p over i. Then all the* A_i *are nonempty parts of A; every element of A is in exactly one of the* A_i.

By way of explanation, note that every (not necessarily surjective) mapping with domain A gives rise to a partition of A, as described in the proposition, by restricting consideration to *those* elements of the codomain at which the fiber is nonempty. In other words, a general mapping specifies more information than just the partition of its domain in that it specifies also the size of the part of the codomain where the fibers are empty. By restricting to the case where the latter part is empty (i.e., to the case of *surjective* mappings), we are in the case of mappings that specify *no more* (and no less) than a partition of the domain. This "justifies" Definition 4.19.

(A similar discussion could have been given for the dual concept "part of *B*". That is, *any* mapping with codomain *B* gives rise to a part of *B*, namely the part whose members are just those elements of *B* that actually are values of the given mapping. However, a general (not-necessarily-injective) mapping specifies much more information than just this "image" part since the members of the image will have some "multiplicity," i.e. will be values of the map at more than one element of the domain (multiplicity = size of fiber). *Injective* mappings (where all nonempty fibers have multiplicity exactly one) are just *parts*, without any additional information.

Here is a typical cograph picture of a partition of a set *A* into three parts:

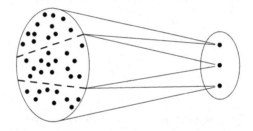

In terms of partitions, the axiom of choice can be rephrased by saying that for *any* partition *p* of any set, there exists at least one choice function *s* that chooses an element from each cell of the partition

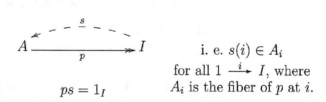

i. e. $s(i) \in A_i$
for all $1 \xrightarrow{i} I$, where
A_i is the fiber of p at i.

The axiom of choice seems quite plausible because we are allowing "arbitrary" mappings *s*. In categories in which only geometrically "reasonable" mappings are allowed, the axiom of choice is usually not true; but this only points out that such categories are *distinct* from the category of constant sets and arbitrary maps, which itself exists as an especially simple extreme case with which to contrast the others (Cantor's abstraction). Some of the opposition to the axiom of choice stems from its consequence, the Banach–Tarski Paradox, which states that a solid ball can be shattered into five parts of such an unreasonable "arbitrary" nature that they can be reassembled into *two* solid balls of the same size as the original. The abstractness of the sets, correlated with the arbitrary nature of the mappings, makes such paradoxes possible, but of course such paradoxes are not possible in the real world where things have variation and cohesion and mappings are *not* arbitrary. Nonetheless, since Cantor mathematicians have found the constant noncohesive sets useful as an extreme case with which continuously variable sets can be contrasted.

Closely related to the notion of partition is the concept of *equivalence relation*.

Definition 4.21: *A relation R from X to Y is a part $R \xrightarrow{\langle p_0, p_1 \rangle} X \times Y$ of the product of X and Y. If $X = Y$, we speak of a relation on X. The* **opposite** *of a binary relation R is the binary relation from Y to X with projections p_1 and p_0. It is denoted R^{op}.*

Notice that there is an abuse of language in omitting the projections involved in defining the binary relation R – and as a mapping R^{op} has the same domain as R!

Exercise 4.22
R^{op} actually is a part of $Y \times X$.

Hint: Use the "twist" map $\tau_{X,Y}$ $\Diamond$

Suppose that $A \xrightarrow{p} I$ is a partition, and let the following diagram be a pullback:

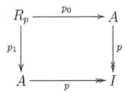

Exercise 4.23
The resulting arrow $R_p \xrightarrow{\langle p_0, p_1 \rangle} A \times A$ is a monomorphism. $\Diamond$

Thus, R_p is a relation on A. It has some special properties we are about to explore, but notice that the proof that $\langle p_0, p_1 \rangle$ is monic *does not depend on the fact that p is epi.* We will need that only when we come to the proof of Proposition 4.32 below. With A as test object and the identity mappings to the vertices of the pullback above we obtain a mapping $r : A \longrightarrow R_p$:

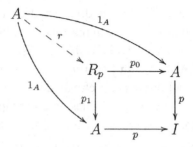

The mapping r satisfies $p_0 r = 1_A = p_1 r$. In general,

Definition 4.24: *A relation* $R \xrightarrow{\langle p_0, p_1 \rangle} X \times X$ *on* X *is called* **reflexive** *iff there is a mapping* $r : X \longrightarrow R$ *such that* $p_0 r = 1_X = p_1 r$.

Exercise 4.25

Recall the diagonal map $X \xrightarrow{\delta_X} X \times X$. A relation R is reflexive if and only if it contains the diagonal part, or if and only if for any (generalized) element x of X we have $\langle x, x \rangle$ is a member of R. $\diamond$

Combining the fact that $\langle p_0, p_1 \rangle$ is monic with uniqueness of mappings to products, we see that for a reflexive relation, r is unique. Thus, a relation either is or is not reflexive. In Section 10.4 we will study *reflexive graphs* determined by two sets E and V and three mappings: $s, t : E \longrightarrow V$ and $r : V \longrightarrow E$ such that $sr = 1_V = tr$. In that case r is part of the reflexive graph structure and need not be unique because $\langle s, t \rangle$ need not be mono.

The relation R_p has two additional important properties we now define.

Definition 4.26: *A relation* $R \xrightarrow{\langle p_0, p_1 \rangle} X \times X$ *on* X *is called* **symmetric** *iff* $R^{\mathrm{op}} \subseteq_{X \times X} R$.

Exercise 4.27

A relation R is symmetric if and only if for any (generalized) element $\langle x, y \rangle$ of $X \times X$ we have

$$\langle x, y \rangle \in R \implies \langle y, x \rangle \in R$$

if and only if the restriction of the twist mapping $\tau_{X,X}$ to R is contained in R. $\diamond$

Exercise 4.28

The relation R_p is symmetric. $\diamond$

A third important property of R_p is the following:

Definition 4.29: *A relation R is called* **transitive** *if and only if for any (generalized) elements $\langle x, y \rangle$ and $\langle y, z \rangle$ of $X \times X$ we have*

$$\langle x, y \rangle \in R \,\&\, \langle y, z \rangle \in R \implies \langle x, z \rangle \in R$$

Exercise 4.30
The relation R_p associated to a map $A \xrightarrow{p} I$ is transitive. ◊

Combining these properties of R_p, we make the following definition:

Definition 4.31: *A relation R on X is called an* **equivalence relation** *iff it is reflexive, symmetric, and transitive.*

Now if we start with an equivalence relation R on X, we may form the coequalizer of $R \underset{p_1}{\overset{p_0}{\rightrightarrows}} X$, which we denote by $X \xrightarrow{p_R} P_R$. This is a partition of X. (We obtain a partition of X by forming the coequalizer of *any* two mappings with codomain X – it is the special properties of equivalence relations that allow the next result.) Taking the equivalence relation of a partition of X and taking the partition from an equivalence relation on X are inverse processes:

Proposition 4.32: *If p is a partition of X, then $p = p_{R_p}$. If R is an equivalence relation on X then $R = R_{p_R}$.*

Exercise 4.33
Prove Proposition 4.32. ◊

4.5 Split Images

There is a version of the choice axiom that does not assume the given map is surjective and correspondingly produces another map that is in general somewhat less than a section.

If $X \xrightarrow{f} Y$ is any mapping, provided X is nonempty, there
exists $Y \xrightarrow{g} X$ for which $fgf = f$.

Exercise 4.34
Show that the principle just enunciated is equivalent to the axiom of choice. (What do you need to assume about the category to prove this equivalence?) ◊

Proposition 4.35: *If $fgf = f$ and f is epimorphic, then $fg = 1_Y$ (i.e., g is a section for f).*

Proof: $fgf = 1_Y f$ and f is right-cancellable.   ∎

Proposition 4.36: *If $fgf = f$ and f is monomorphic, then $gf = 1_X$ (i.e., g is a retraction for f).*

Proof: Exercise.   ∎

It will be recalled that we proved earlier that if $X \xrightarrow{f} Y$ is monic and $X \neq 0$, then f has a retraction using only the existence of complements for all parts and that any nonempty set has an element; thus, this monic case of the $fgf = f$ principle does not involve "choice" functions. The $fgf = f$ principle can be used to derive the main properties of the **image factorization** of a mapping f (which we illustrate with external and internal diagrams)

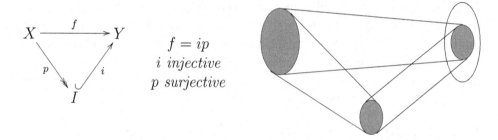

$$f = ip$$
$$i \ injective$$
$$p \ surjective$$

for any f by strengthening it to a split image (meaning p has a splitting or section s):

$X \xrightarrow{f} Y$

s, p, i, I

$f = ip$
i injective
$ps = 1_I$

Exercise 4.37
If the factorization $f = ip$ is a split image for f, then i is the part of y whose members are exactly all those elements of Y that are values of f; i.e.,

$$\forall T \xrightarrow{y} Y[y \in i \iff \exists x[y = fx]]$$

◇

The section s in the split-image concept is *not* determined by f (even though the image factorization $f = ip$ is essentially determined), but on the other hand it makes some calculations somewhat easier. The next few exercises are concerned with showing how split images can be derived from the pseudosections g for arbitrary nonempty X.

Exercise 4.38

If $fgf = f$, we can explicitly define (by composition) another "better" $\bar{g}$ for which both

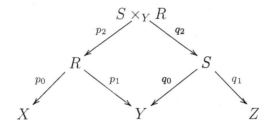

$$fgf = f$$
$$\text{and}$$
$$\bar{g}f\bar{g} = \bar{g}$$

The improvement process goes no further: $\bar{\bar{g}} = \bar{g}$; i.e. if g already satisfies *both* equations, the process ($^-$) will not change it. $\Diamond$

Exercise 4.39

If $f = ip$, where p has a section s and i has a retraction r, and if we define $g \overset{\text{def}}{=} sr$, then the resulting g will already satisfy both $fgf = f$ and $gfg = g$. $\Diamond$

Exercise 4.40

Suppose $fgf = f$ and let i be an equalizer of fg with 1_Y, where Y is the codomain of f. Show, using the universal mapping property of equalizers, that a unique map p exists for which $f = ip$. Show that $pgi = 1_E$ and hence that if we define $s \overset{\text{def}}{=} gi, r \overset{\text{def}}{=} pg$ we get that $ps = 1_I$ and $ri = 1_E$. Moreover $ir = fg$ and $sp = gf$. How must g be related to f in order to have further that $g = sr$? $\Diamond$

Although most categories do not have split images, often they do have image factorization $f = ip$ in the sense that i is the smallest part of the codomain through which f factors and p is the proof that f belongs to i. (It then follows that p is an epimap.)

With images available we can define a composition of relations and give another characterization of transitive relations.

If $R \xrightarrow{\langle p_0, p_1 \rangle} X \times Y$ is a relation from X to Y and $S \xrightarrow{\langle q_0, q_1 \rangle} Y \times Z$ is a relation from Y to Z, we denote by $S \times_Y R$ the pullback of p_1 and q_0; thus, we have

Definition 4.41: *The* **relational composite** *of R and S is the image*
$S \circ R = I_{\langle p_0 p_2, q_1 q_2 \rangle}$ *of the mapping* $\langle p_0 p_2, q_1 q_2 \rangle : S \times_Y R \longrightarrow X \times Z.$

Note that the graph γ_f of a mapping $X \xrightarrow{\ f\ } Y$ is a relation from X to Y.

Exercise 4.42
If $X \xrightarrow{\ f\ } Y$ and $Y \xrightarrow{\ g\ } Z$ then $\gamma_{gf} = \gamma_g \circ \gamma_f.$ ◊

Exercise 4.43
A relation $R \xrightarrow{\langle p_0, p_1 \rangle} X \times X$ on X is transitive iff $R \circ R \subseteq_{X \times X} R.$ ◊

4.6 The Axiom of Choice as the Distinguishing Property of Constant/Random Sets

Much of mathematics revolves around determining whether specific maps f have sections s or not. In a general way this is because the sets of mathematical interest have some variation/cohesion within them and the maps (such as s) in the corresponding categories are accordingly not "arbitrary". However, for the category S of constant sets and arbitrary maps, the axiom of choice is usually assumed to hold.

The name "choice" comes from the following observation: A map $X \xrightarrow{\ f\ } Y$ can be viewed as a family of sets

$$X_y = \{x \mid fx = y\}$$

parameterized by the elements y of Y. Then a section s for f is a single rule that chooses one element from each set in the family:

$$fs = 1_Y \implies f(sy) = y \implies s(y) \in X_y \text{ for each } y$$

The notion of part (including membership in a part and part defined by a condition) as just used in the explanation of the word "choice," can itself be entirely explained in terms of composition of mappings (without assuming rigid membership chains or imposing an a priori model for every part).

The axiom of choice implies two other properties (which will not both be true of genuinely variable sets), both of which can be stated in terms of parts. The **Boolean property** states that for any part $i_0 : A_0 \hookrightarrow X$ of any set X there is another part $i_1 : A_1 \hookrightarrow X$ of X (the **complement** of A_0) such that X is the coproduct of A_0 and A_1; i.e. for all Y, for any pair of arrows $f_0 : A \longrightarrow Y$, $f_1 : A_1 \longrightarrow Y$ with codomain Y there exists a unique $f : X \longrightarrow Y$, making the diagram below commutative:

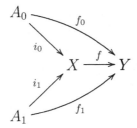

It can be proved that the Boolean property follows from the axiom of choice. Now we only point out the randomness implied by the Boolean property: The maps f_0, f_1 can be specified independently (no compatibility required on any "boundary" between A_0, A_1) and still be part of a single map (on the whole X) that the Boolean category allows; moreover, the uniqueness of the extended map f shows that A_0, A_1 together exhaust X (for if there were any additional part, f could be varied on it without disturbing the condition that f restricts to the given f_0, f_1).

There are many important categories of variable sets that satisfy the Boolean property without satisfying the axiom of choice. The most important such examples are determined by groups, and in Chapter 10 we will study enough group theory to understand in a general way how these examples operate.

The other very restrictive consequence of the axiom of choice is sometimes called the *localic property* or the property that "parts of 1 separate". Recall from Exercise 2.44 that in $\mathcal{S}/X$ the terminal object has many parts in general, and indeed they separate maps in that category. In all of our categories there turn out to be suitable objects B such that "parts of B separate," meaning that there are enough elements whose domains are (domains of) parts of B to distinguish maps between any X, Y:

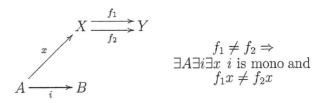

A category of sets is called **localic** if this holds for $B = 1$.

It can be shown that the localic property follows from the axiom of choice. Although most of the examples of categories of variable sets are not localic, there are a great many that are localic even without satisfying the axiom of choice. The most important localic examples correspond to topological spaces. In Exercise 10.27 we will learn enough about topological spaces to comprehend how sets varying over a topological parameter space "restrict" from one open region to a smaller one.

There are a great many groups and a great many topological spaces, and each one of each gives an example of a category of variable sets. But examples that are both localic and Boolean and yet do not satisfy the axiom of choice are harder to come by and indeed have only been known since Cohen's 1963 discoveries. The study of such "independence proofs" is a part of the multifaceted lore of variable sets.

4.7 Additional Exercises

Exercise 4.44
Prove Proposition 4.20.

Exercise 4.45

(a) Show that the category of finite-dimensional vector spaces has finite colimits.

(b) Show that in the category of finite-dimensional vector spaces epimorphisms are surjective linear transformations and, moreover, any epimorphism has a section.

Exercise 4.46

(a) Show that in a slice category S/X a morphism h from $f : A \longrightarrow X$ to $g : B \longrightarrow X$ is an epimorphism iff h is a surjective mapping in S.

(b) Show that the slice categories S/X have finite colimits.

   Hint: They are computed using colimits in S.

Exercise 4.47

(a) Show that epimorphisms in the category of M-sets are equivariant mappings that are surjective.

(b) Show that these categories have finite colimits.

   Hint: They are computed using colimits in S.

Exercise 4.48

(a) Show that the category of partially ordered sets has finite colimits.

(b) Show that epimorphisms in the category of partially ordered sets are surjective (order-preserving) mappings.

Exercise 4.49

Describe equivalence relations in the categories $\mathcal{S}/X$.

Exercise 4.50

Describe equivalence relations in the categories of M-sets.

Exercise 4.51

(a) Describe equivalence relations in the category of groups. In the category of groups equivalence relations are called **congruences.** The congruences on a group correspond bijectively to the normal subgroups.

(b) Show that the analogue of Proposition 4.20 holds in the category of groups (where it is usually called the First Isomorphism theorem.)

Exercise 4.52

Show that the axiom of choice holds in the categories $\mathcal{S}/X$.

Exercise 4.53

Show that the axiom of choice does *not* hold in the category of groups.

Exercise 4.54

Show that the axiom of choice does *not* hold in general in the category of M-sets.

Hint: There is a very small counterexample using the action of a one-letter alphabet on sets with at most two elements. In fact, it can be shown that the axiom of choice fails in M-sets unless M is the monoid with only one element (in which case M-sets are equivalent to $\mathcal{S}$).

Exercise 4.55

On the other hand, if a monoid has an inverse for every element (i.e., is a group G), then as mentioned in Section 4.6 the Boolean property holds for G-sets.

Hint: Consider the nature of monomorphisms in G-sets; they are rather special.

5

Mapping Sets and Exponentials

5.1 Natural Bijection and Functoriality

The essential properties of the product operation can be summed up by the figure

$$\boxed{P} \qquad \frac{X \longrightarrow Y_0 \times Y_1}{X \longrightarrow Y_0, \, X \longrightarrow Y_1} \quad \downarrow\uparrow$$

where the horizontal bar will be interpreted in such contexts to mean there is a natural process that, to every arrow of the type indicated above the bar, assigns one of the type indicated below the bar, and there is also a natural process from below to above, and these two natural processes are inverse to each other in the sense that following one process by the other gives nothing new. Of course, in the example of products the two processes in question are supposed to be

(1) taking the components of a map whose codomain is a product, and
(2) "pairing."

Just from the idea that there should be such a couple of inverse processes, we can (by a "bootstrap" procedure) discover more specifically how the processes must work. Namely, from the presumption that $\boxed{P}$ should function for all X, Y_0, Y_1, and for all maps as indicated, we can deduce two special cases: Let Y_0, Y_1 be arbitrary but suppose $X = Y_0 \times Y_1$; then, we make the very special choice "above the bar" of $1_{Y_0 \times Y_1}$ which will correspond to something specific, namely, the projections below the bar

$$\frac{Y_0 \times Y_1 \xrightarrow{\ 1\ } Y_0 \times Y_1}{Y_0 \times Y_1 \xrightarrow{\ p_0\ } Y_0, \, Y_0 \times Y_1 \xrightarrow{\ p_1\ } Y_1}$$

Returning to the general X, the same projections are the means by which, through composition, the general top-to-bottom process is effected as follows:

$$\frac{X \xrightarrow{f} Y_0 \times Y_1}{X \xrightarrow{p_0 f} Y_0, \; X \xrightarrow{p_1 f} Y_1} \quad \downarrow$$

Of course the bottom-to-top process is usually just indicated by pairing (the result of which is often denoted by brackets),

$$\frac{X \xrightarrow{\langle f_0, f_1 \rangle} Y_0 \times Y_1}{X \xrightarrow{f_0} Y_0, \; X \xrightarrow{f_1} Y_1} \quad \uparrow$$

but this can also be analyzed further: First consider the special case in which X is arbitrary but both $Y_0 = X$ and $Y_1 = X$; then, we have the possibility of considering the very special pair of maps below the bar, which by the pairing process will lead to a specific map

$$\frac{X \xrightarrow{\delta_X} X \times X}{X \xrightarrow{1_X} X, \; X \xrightarrow{1_X} X}$$

known as the diagonal $\delta_X = \langle 1_X, 1_X \rangle$. The diagonal helps via composition to effect the general case of the pairing process provided one first develops the "functoriality" of product (which can also be regarded as deduced from the basic $\boxed{\text{P}}$):

If $X_0 \xrightarrow{f_0} Y_0, \; X_1 \xrightarrow{f_1} Y_1$ are *any* two mappings, there is an induced map

$$X_0 \times X_1 \xrightarrow{f_0 \times f_1} Y_0 \times Y_1$$

called the (Cartesian) product of the two maps that is characterized by the equations

$$
\begin{array}{ccc}
X_0 \times X_1 & \xrightarrow{f_0 \times f_1} & Y_0 \times Y_1 \\
\Big\downarrow{\scriptstyle p_k} & & \Big\downarrow{\scriptstyle p_k} \\
X_k & \xrightarrow{f_k} & Y_k
\end{array}
\qquad
\begin{array}{c}
p_k(f_0 \times f_1) = f_k p_k \\
k = 0, 1
\end{array}
$$

or, in terms of values, by

$$(f_0 \times f_1)\langle x_0, x_1 \rangle = \langle f_0 x_0, f_1 x_1 \rangle$$

Exercise 5.1

("Functoriality of product") If $Y_0 \xrightarrow{g_0} Z_0$, $Y_1 \xrightarrow{g_1} Z_1$ are further maps, then

$$(g_0 \times g_1)(f_0 \times f_1) = (g_0 f_0) \times (g_1 f_1)$$

$\lozenge$

Now the analysis of the pairing in terms of the product of maps and composition with δ_X is just this:

If $X_0 = X$, $X_1 = X$ and if $X \xrightarrow{f_0} Y_0$, $X \xrightarrow{f_1} Y_1$ then

$$\langle f_0, f_1 \rangle = (f_0 \times f_1)\delta_X$$

as is easily checked by following both sides of the equation with the projections.

This two-level method, applied here to summarize the transformations possible with the product construction, will be applied to many different constructions later, so let us restate the levels: The *crude idea* that there should be a natural invertible process $\downarrow\uparrow$ between maps of two kinds is refined to a *precise suggestion* of how the process can be carried out, namely, to apply a functorial construction and compose with a specific map; a relationship between the two levels is essentially that the crude idea of the invertible process, applied to a very special case and to the very special identity map, yields that specific map, which can be used in composition to help effect the process.

5.2 Exponentiation

We will now discuss another important construction, that of *exponentiation* or *mapping set,* whose description involves a strictly analogous two-tier process for objects X, Y and B:

$$\frac{X \longrightarrow Y^B}{X \times B \longrightarrow Y} \quad \downarrow\uparrow$$

Immediately we can derive from this crude idea what the elements of Y^B must be by considering the special case $X = 1$: The process

$$\frac{1 \longrightarrow Y^B}{B \longrightarrow Y} \quad \downarrow$$

must be an invertible process (recall from Exercise 3.22 that $1 \times B \cong B$), i.e. there must be just as many elements of Y^B as there are actual maps $B \longrightarrow Y$.

Now if we are given any map

$$X \times B \xrightarrow{f} Y$$

whose domain is equipped with a product structure (frequently one refers to such f as a function of two variables), then for each element $1 \xrightarrow{x} X$ of X we can consider (with $\bar{x}$ the constant composite: $B \longrightarrow 1 \xrightarrow{x} X$)

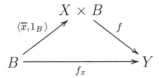

In other words f_x is the map $B \longrightarrow Y$ whose value at any element b is

$$f_x(b) = f(x, b)$$

That is, such a single map f on a product $X \times B$ gives rise to a family, parameterized by X, of maps $B \longrightarrow Y$. Depending on the size of X and on the nature of f, one may have (but usually will not have) the following two properties:

(1) Every map $B \longrightarrow Y$ occurs as f_x for at least one $1 \xrightarrow{x} X$ (i.e., every $B \longrightarrow Y$ has at least one "name" provided by the scheme X, f).
(2) $f_{x_1} = f_{x_2}$ only when $x_1 = x_2$, (i.e., "nameable" maps $B \longrightarrow Y$ have only one "name".)

In case both properties are true, one often writes Y^B in place of X and calls the map $Y^B \times B \longrightarrow Y$ *evaluation* (rather than calling it f). For any $B \xrightarrow{f} Y$ let $1 \xrightarrow{\ulcorner f \urcorner} Y^B$ be the unique element of Y^B guaranteed by the properties (1) and (2); then we have for all f, b

$$
\begin{array}{ccc}
1 & \xrightarrow{\;\;b\;\;} & B \\
{\scriptstyle \langle \ulcorner f \urcorner, b \rangle} \downarrow & & \downarrow {\scriptstyle f} \qquad \mathrm{eval}\langle \ulcorner f \urcorner, b \rangle = fb \\
Y^B \times B & \xrightarrow[\mathrm{eval}]{} & Y
\end{array}
$$

Such a universal map-parameterizing scheme (one which does enjoy both properties (1) and (2)) will have a unique relationship to *any* map-parameterizing scheme X, f (in which $X \times B \xrightarrow{f} Y$ and any $1 \xrightarrow{x} X$ names $B \xrightarrow{f_x} X$), as expressed by the natural invertible process

$$\frac{X \xrightarrow{\ulcorner f \urcorner} Y^B}{X \times B \xrightarrow{f} Y} \;\; \downarrow\uparrow$$

namely,

$$\ulcorner f \urcorner x = \ulcorner f_x \urcorner$$

Thus, this extended naming process is related to the evaluation map by

$$\mathrm{eval}(\ulcorner f \urcorner x, b) = f(x, b)$$

or

$$\text{eval}(\ulcorner f \urcorner \times 1_B) = f$$

For any f, $\ulcorner f \urcorner$ is the only map for which the latter equation is true, and thus we can say briefly that any f is uniquely derivable from the evaluation map.

As an example imagine that the elements of B represent particles of a continuous body of matter (such as a cloud); also imagine that a set E represents the points of ordinary space and a set T represents the instants of time within a certain interval. Then,

$$E^B = \text{placements of } B$$

is a set, *each element* of which determines an entire map $B \longrightarrow E$ telling where in E each particle of B is (i.e., a *placement* of B in E). Thus, a *motion* of B during the time interval T could be considered as a mapping

$$T \xrightarrow{\ulcorner m \urcorner} E^B$$

whose value at any instant is a placement.

By the fundamental transformation law for mapping sets, $\ulcorner m \urcorner$ corresponds to a unique

$$T \times B \xrightarrow{m} E$$

which represents the same motion in a different mathematical aspect: for each instant t and particle b, the value $m(t, b)$ is the point in space at which the particle b finds itself at instant t during the motion. But there is still a third way to describe the same motion. Because of the natural twist map $B \times T \xrightarrow{\tau} T \times B$, we have $B \times T \xrightarrow{m\tau} E$, and hence by a different instance of the fundamental transformation we find

$$B \xrightarrow{\ulcorner m\tau \urcorner} E^T$$

which seems to describe the motion equally well. Here

$$E^T = \text{paths in } E$$

is a set, each element of which determines a whole map $T \longrightarrow E$ from instants of time to points of space (i.e., a "path"). The given motion determines, for each particle b, the whole path that b follows during T. The expressions

$$\frac{\dfrac{B \longrightarrow E^T}{B \times T \longrightarrow E}}{T \longrightarrow E^B}$$

are thus the three descriptions of the same motion of the cloud. The first and last involve the infinite-dimensional function spaces E^T and E^B, whereas the middle one involves only the finite dimensional $B \times T \xrightarrow{\sim} T \times B$.

Although maps whose *codomain* is a function space (= mapping set or exponential) can always be "transformed down" in the indicated manner, nothing of the sort is true in general for mappings whose *domain* is a function space. Mappings whose domain is a function space are often called

Operators or Functionals

and include such things as differentiation and integration. An example of an integration functional arises when we consider the mass distribution on the body B and the resulting map

$$E^B \longrightarrow E$$

whose value at any placement is the *center of mass* (which is a point of E) of that placement. Thus, the description of a motion by means of

$$T \longrightarrow E^B$$

is necessary if we want to compute by composition

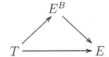

the map whose typical value is the instantaneous position of the center of mass. On the other hand, a typical differentiation operator is the "velocity"

$$E^T \longrightarrow V^T$$

whose value at any path of points is the corresponding path of velocity vectors. If we are to calculate the velocity field resulting from a particular motion m of an extended body B, the appropriate description of the motion will be

$$B \longrightarrow E^T$$

for then we can just calculate by composition

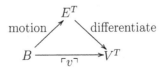

and then apply once again the fundamental transformation to get $\ulcorner v\tau \urcorner$

$$T \longrightarrow V^B$$

whose value at any instant is the *velocity field* over the body.

5.3 Functoriality of Function Spaces

We will take as an axiom the following:

AXIOM: EXPONENTIATION AXIOM

For any two sets B, Y there exists a set Y^B and an evaluation $Y^B \times B \xrightarrow{\text{eval}} Y$ having the universal property that for each X and for each $X \times B \xrightarrow{f} Y$ there is a unique $X \xrightarrow{\ulcorner f \urcorner} Y^B$ for which eval $(\ulcorner f \urcorner \times 1_B) = f$. Briefly,

$$\frac{X \xrightarrow{\ulcorner f \urcorner} Y^B}{X \times B \xrightarrow{f} Y}$$

with the evaluation map arising as the unique f for which $\ulcorner f \urcorner = 1_{Y^B}$.

Proposition 5.2: *If $Y_1 \xrightarrow{\varphi} Y_2$ then there is a unique $Y_1^B \xrightarrow{\varphi^B} Y_2^B$ for which*

$$\varphi^B \ulcorner g \urcorner = \ulcorner \varphi g \urcorner \qquad B \xrightarrow{g} Y_1 \xrightarrow{\varphi} Y_2$$

for all g.

Proof: The axiom says that to construct a map whose codomain is a function space Y_2^B it suffices to construct the corresponding map whose domain is a product, in our case $Y_1^B \times B$. But this is easily done as the following composition:

$$\frac{Y_1^B \times B \xrightarrow{\text{eval}_1} Y_1 \xrightarrow{\varphi} Y_2}{Y_1^B \xrightarrow{\varphi^B} Y_2^B}$$

Notice that this means that $\varphi^B = \ulcorner \varphi \, \text{eval}_1 \urcorner$ by the axiom, and thus

$$\text{eval}_2(\varphi^B \times 1_B) = \varphi \, \text{eval}_1$$

Furthermore $(\varphi^B \times 1_B)(\ulcorner g \urcorner \times 1_B) = \varphi^B \ulcorner g \urcorner \times 1_B$. Thus,

$$\begin{aligned}
\text{eval}_2(\varphi^B \ulcorner g \urcorner \times 1_B) &= \text{eval}_2(\varphi^B \times 1_B)(\ulcorner g \urcorner \times 1_B) \\
&= \varphi \, \text{eval}_1(\ulcorner g \urcorner \times 1_B) = \varphi g \\
&= \text{eval}_2(\ulcorner \varphi g \urcorner \times 1_B)
\end{aligned}$$

That is, both $\varphi^B \ulcorner g \urcorner$ and $\ulcorner \varphi g \urcorner$ give the same result when crossed with 1_B and then composed with eval_2. Since the axiom states the *uniqueness* of maps yielding a given result under this two-step transformation, we conclude that they are equal, as Proposition 5.2 called for. ∎

Remark: Note that the argument works just as well if $\ulcorner g \urcorner$ is a *generalized* element of Y_1^B, that is, for $X \times B \xrightarrow{g} Y_1$.

Exercise 5.3
If $Y_1 \xrightarrow{\varphi} Y_2 \xrightarrow{\psi} Y_3$, then

$$(\psi\varphi)^B = \psi^B \varphi^B$$

◊

The preceding exercise establishes the "covariant functoriality of induced maps on function spaces" with given domain B. But if we fix the codomain space Y and instead let the *domain* space B vary along maps, we find that there are again induced maps on the function spaces, but of a *contravariant* nature, in the following sense:

Proposition 5.4: *If $B_2 \xrightarrow{\beta} B_1$ then there is a unique*

$$Y^{B_1} \xrightarrow{Y^\beta} Y^{B_2}$$

for which

$$Y^\beta \ulcorner g \urcorner = \ulcorner g\beta \urcorner$$

for all g.

Proof: Again we construct by composition the map $Y^{B_1} \times B_2 \longrightarrow Y$ that uniquely corresponds by the fundamental transformation of the axiom to the Y^β desired:

$$Y^{B_1} \times B_2 \xrightarrow{1 \times \beta} Y^{B_1} \times B_1 \xrightarrow{\mathrm{eval}_1} Y$$

This can be done along the lines of the proof of Proposition 5.2. ∎

Proposition 5.5: *If $B_3 \xrightarrow{\alpha} B_2 \xrightarrow{\beta} B_1$, then*

$$Y^{\beta\alpha} = Y^\alpha Y^\beta$$

Proof: The proof of the proposition is simply a "higher" expression of the associativity of composition in

$$B_3 \xrightarrow{\alpha} B_2 \xrightarrow{\beta} B_1$$
$$\downarrow g$$
$$Y$$

One way $Y^\beta \ulcorner g \urcorner = \ulcorner g\beta \urcorner$; hence,

$$Y^\alpha Y^\beta \ulcorner g \urcorner = Y^\alpha \ulcorner g\beta \urcorner = \ulcorner (g\beta)\alpha \urcorner$$

but the other way

$$Y^{\beta\alpha} \ulcorner g \urcorner = \ulcorner g(\beta\alpha) \urcorner$$

hence the two are equal for each g, which accounts for the conclusion. ∎

In case $B_2 \xrightarrow{\beta} B_1$ is injective, then $Y^{B_1} \xrightarrow{Y^\beta} Y^{B_2}$ is often called the operator of *restriction* (of maps, restriction to the smaller domain along the part of their original domain).

The next proposition shows that the operation on mappings that we take as most fundamental, namely composition, can itself be expressed via an actual mapping – at least when the three sets involved are fixed.

Proposition 5.6: *There is a map* $Y_2^{Y_1} \times Y_1^B \xrightarrow{\ulcorner c \urcorner} Y_2^B$ *such that*

$$\ulcorner c \urcorner (\ulcorner \varphi \urcorner, \ulcorner g \urcorner) = \ulcorner \varphi g \urcorner$$

for all

$$B \xrightarrow{g} Y_1 \xrightarrow{\varphi} Y_2$$

Proof: Define c by:

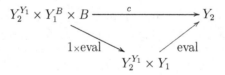

∎

Note that $\ulcorner c \urcorner$ can be further transposed by another instance of the axiom to give

$$Y_2^{Y_1} \longrightarrow (Y_2^B)^{(Y_1^B)}$$

whose value at any $\ulcorner \varphi \urcorner$ is the name $\ulcorner \varphi^B \urcorner$ of the (covariantly) induced map on the function spaces with domain (= exponent) B. That is, the inducing operation is itself represented by a mapping. In a similar fashion (taking a differently labeled instance of c) the third way (as in our example of motion) yields

$$B_1^{B_2} \longrightarrow (Y^{B_2})^{(Y^{B_1})}$$

representing the contravariant-inducing process by an actual map.

Recalling that the set 2 represents, via the notion of characteristic function, the arbitrary parts of an arbitrary set B,

$$\frac{B \longrightarrow 2}{? \lhook\joinrel\longrightarrow B}$$

we can combine this bijection with the function-space axiom to get

$$\frac{1 \longrightarrow 2^B}{? \lhook\joinrel\longrightarrow B}$$

so that the elements of the set 2^B serve as effective names for the arbitrary parts of B and, in particular, the "number" of parts can be defined to be the number of elements of 2^B. Moreover, the meaning of the evaluation map in this case is just "membership":

$$2^B \times B \xrightarrow{\text{eval}} 2$$

$$\text{eval}(\ulcorner \varphi \urcorner, b) = \varphi b$$

thus, if φ is the characteristic function of a part i, we could define

$$[b \in i] = \text{eval}(\ulcorner \varphi \urcorner, b)$$

as the truth value of the statement that b is a member of i.

Exercise 5.7
If $B_2 \xrightarrow{\beta} B_1$, then

$$2^{B_1} \xrightarrow{2^\beta} 2^{B_2}$$

represents by a map the operation of taking the *inverse image* along β of arbitrary parts of B_1. ◊

To justify the use of the exponential notation for function spaces, we will show that for finite sets the number of maps from B to Y is the numerical exponential – number of elements of Y to the power number of elements of B.

Proposition 5.8:

$$Y^0 \xrightarrow{\sim} 1$$

$$Y^{A+B} \xrightarrow{\sim} Y^A \times Y^B$$

Proof: To construct $1 \longrightarrow Y^0$ is equivalent to constructing $1 \times 0 \longrightarrow Y$, but since $0 \times 1 \xrightarrow{\sim} 1 \times 0$, that is equivalent to constructing $0 \longrightarrow Y^1$, which is *unique*;

the composite $1 \longrightarrow Y^0 \longrightarrow 1$ is the identity because maps to 1 are unique, and the composite $Y^0 \longrightarrow 1 \longrightarrow Y^0$ must be the identity since it corresponds to a map $0 \longrightarrow (Y)^{(Y^0)}$ of which there is only one. For the second statement from the proposition consider the injections i_A, i_B into the sum $A + B$; by the functoriality of $Y^{(\ )}$ these induce "restriction" maps $Y^{A+B} \longrightarrow Y^A$, $Y^{A+B} \longrightarrow Y^B$, which can be paired to yield

$$Y^{A+B} \xrightarrow{\ \langle Y^{i_A}, Y^{i_B} \rangle\ } Y^A \times Y^B$$

the map mentioned in the statement, which we want to show is invertible. To construct an inverse we must construct first a candidate

$$Y^A \times Y^B \longrightarrow Y^{A+B}$$

which is equivalent to constructing

$$A + B \longrightarrow Y^{(Y^A \times Y^B)}$$

which must necessarily be the "copair" of two maps

$$A \longrightarrow Y^{(Y^A \times Y^B)}$$
$$B \longrightarrow Y^{(Y^A \times Y^B)}$$

But such a couple of maps is equivalent to

$$Y^A \times Y^B \longrightarrow Y^A$$
$$Y^A \times Y^B \longrightarrow Y^B$$

for which the obvious choices are the *projections*; tracing back through the equivalences, we then have our map

$$Y^A \times Y^B \longrightarrow Y^{A+B}$$

which we must show is the inverse. The map is the one whose value at a pair of names is the name of the copair,

$$\langle \ulcorner f_A \urcorner, \ulcorner f_B \urcorner \rangle \qquad \text{goes to} \qquad \ulcorner \begin{cases} f_A \\ f_B \end{cases} \urcorner$$

which is indeed clearly inverse to the previous one whose value at $\ulcorner f \urcorner$ is $\langle \ulcorner f i_A \urcorner, \ulcorner f i_B \urcorner \rangle$. $\blacksquare$

Notice that the effect of Proposition 5.8 is to represent the invertible process characteristic of sum

$$\frac{A + B \longrightarrow Y}{A \longrightarrow Y, B \longrightarrow Y} \ \downarrow\uparrow$$

as an actual invertible *map*. In a similar way the processes characteristic of product, and even the processes characteristic of exponentiation itself (!), can be represented by actual invertible *maps* by allowing the exponential sets to intervene, as will be shown in the following two propositions.

Proposition 5.9:

$$1^B \xrightarrow{\sim} 1$$

$$(Y_0 \times Y_1)^B \xrightarrow{\sim} Y_0^B \times Y_1^B$$

Proof: Since the assertions of the proposition are that certain specific maps are invertible, we must find these inverses. The inverse of the unique $1^B \longrightarrow 1$ is the "name" $1 \longrightarrow 1^B$ of the unique $B \longrightarrow 1$. The map $\langle \pi_0^B, \pi_1^B \rangle$ has an inverse,

$$Y_0^B \times Y_1^B \longrightarrow (Y_0 \times Y_1)^B$$

which is constructed as the map corresponding to

$$Y_0^B \times Y_1^B \times B \xrightarrow{\epsilon} Y_0 \times Y_1$$

where

$$\epsilon \langle \ulcorner g_0 \urcorner, \ulcorner g_1 \urcorner, b \rangle = \langle g_0 b, g_1 b \rangle$$

∎

Thus Proposition 5.9 internalizes or objectifies the defining property of the terminal object and the defining process of the product concept.

Proposition 5.10:

$$Y \xrightarrow{\sim} Y^1$$

$$Y^{X \times B} \xrightarrow{\sim} (Y^B)^X$$

Proof: The diagrams:

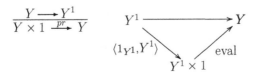

define the two maps inverse to each other for the first "equation". For the second, we have

$$
\frac{\dfrac{Y^{X\times B} \xrightarrow{\ ?\ } (Y^B)^X}{Y^{X\times B}\times X \xrightarrow{\ ?\ } (Y^B)}}{Y^{X\times B}\times X\times B \xrightarrow{\ \mathrm{eval}\ } Y}
\qquad
\frac{\dfrac{(Y^B)^X \xrightarrow{\ ?\ } Y^{X\times B}}{(Y^B)^X\times(X\times B)\xrightarrow{\ ?\ } Y}}{((Y^B)^X\times X)\times B \longrightarrow Y}
$$

$$
((Y^B)^X\times X)\times B \xrightarrow{\qquad\qquad\qquad} Y
$$
$$
\mathrm{eval}_{X\times 1_B}\searrow \qquad \nearrow \mathrm{eval}_B
$$
$$
(Y^B)\times B
$$

Schematically, the first map is $\ulcorner f\urcorner \mapsto \ulcorner x\mapsto f(x,-)\urcorner$, and its inverse is $\ulcorner\varphi\urcorner \mapsto \ulcorner\langle x,b\rangle \mapsto (\varphi x)(b)\urcorner$. (Do Exercise 5.16.) Thus, the invertible *map* of the proposition internalizes the *process*

$$
\frac{X\times B \longrightarrow Y}{X \longrightarrow Y^B}
$$

which is the defining property of exponentiation itself.   ∎

   More surprising is that the existence of exponentiation implies a fundamental "equation" that does not mention exponentiation but is only concerned with the lower-order operations of sum and product. We refer to the *distributive law*, which is discussed in Section 7.2.

5.4 Additional Exercises

Exercise 5.11
In Exercise 3.46 we saw that the linear transformations between finite-dimensional vector spaces V and V' have a natural commutative group operation. In fact they even form a vector space, often denoted $L(V, V')$. This is easily seen by remembering that the linear transformations may be represented by matrices (though this representation depends on choosing bases for V and V').
   Show, however, that $L(V, V')$ does not satisfy the exponential axiom.

Hint: There is a simple dimension argument to show this.

Exercise 5.12
The categories $\mathcal{S}/X$ have mapping sets. To see how these are constructed, let $A \xrightarrow{\alpha} X$ and $B \xrightarrow{\beta} X$ be objects. We can write these as families of fibers: $\langle A_x\rangle_{x\in X}$ and $\langle B_x\rangle_{x\in X}$. The mapping set α^β has, as fiber over the element x, the mapping set in $\mathcal{S}$ denoted $A_x^{B_x}$.
   Show that α^β has the correct universal property.

Exercise 5.13

For any monoid M the category of M-sets (see Exercise 3.51) has mapping sets. Construct these from the equivariant mappings and show that the universal property holds.

Hint: The construction is a little more subtle than might be supposed. However, if M happens to be a group, your general description of the M-set Y^X for M-sets X and Y can be shown to be equivalent to the following simpler description: Consider the set of all arbitrary mappings from the underlying set of X to the underlying set of Y and define on it an appropriate action of M.

(*Caution:* If your definition of this appropriate action does not explicitly involve the inverse operation of the group, you will not be able to prove the universal property.)

Exercise 5.14

The category of partially ordered sets has exponentials. If X and Y are partially ordered sets, the mappings from X to Y, which are order-preserving, have a natural order inherited from the order on Y.

Show that this ordered set has the universal property of the exponential.

Exercise 5.15

Complete the proof of Proposition 5.9 by showing that the maps defined are inverse to each other as claimed.

Exercise 5.16

Complete the proof of Proposition 5.10 by showing that the maps defined are inverse to each other as claimed.

Exercise 5.17

Let $\mathcal{A}$ and $\mathcal{B}$ be categories. The concept of **functor** F from $\mathcal{A}$ to $\mathcal{B}$ is defined in Appendix C.1, as well as in 10.18.

(a) Show that the assignments $\Delta(X) = (X, X)$ for any set X and $\Delta(f) = (f, f)$ for any mapping f define a functor $\Delta : \mathcal{S} \longrightarrow \mathcal{S} \times \mathcal{S}$; Δ is called the **diagonal** functor.

(b) Show that the assignments $(- \times B)(X) = X \times B$ for any set X and $(- \times B)(f) = f \times 1_B$ for any mapping f define a functor $(- \times B) : \mathcal{S} \longrightarrow \mathcal{S}$.

(c) Show that the assignments $(-)^B(X) = X^B$ for any set X and $(-)^B(f) = f^B$ for any mapping f define a functor $(-)^B : \mathcal{S} \longrightarrow \mathcal{S}$.

(d) Show that the assignments $- \times -(X, Y) = X \times Y$ for any pair of sets (X, Y) and $- \times -(f, g) = f \times g$ for any pair of mappings (f, g) define a functor $- \times - : \mathcal{S} \times \mathcal{S} \longrightarrow \mathcal{S}$.

(e) If V and W are vector spaces considered as categories (Exercise 1.31) show that any linear transformation between them is a functor.

(f) If X and Y are partially ordered sets considered as categories (Exercise 1.31), show that a functor between them is the same thing as a monotone mapping.

6

Summary of the Axioms and an Example
of Variable Sets

6.1 Axioms for Abstract Sets and Mappings

We have now seen most of the axioms we will require of the category S of abstract sets and mappings. As we progressed, some of the earlier axioms were included in later axioms. For example, the existence of the one-element set is part of the axiom that S has finite limits. Although we did not insist on it earlier, it is also the case that some of the axioms are more special than others. By this we mean that even though they hold in S they will not generally hold in categories of variable or cohesive sets. Thus, we are going to review the axioms here so that they can be considered all at once and grouped according to their generality.

The very first axiom, of course, is

AXIOM: S IS A CATEGORY
We have been emphasizing all along that the fundamental operation in a category, composition, is the basic tool for both describing and understanding all of the other properties of S.

The next group of three axioms is satisfied by any category of sets, variable or constant. In fact a category satisfying them is called a **topos** (in the **elementary** sense), and these categories have been studied intensively since 1969.

AXIOM: FINITE LIMITS AND COLIMITS
S has all finite limits and colimits.

AXIOM: EXPONENTIATION
There is a mapping set Y^X for any objects X and Y in S.

AXIOM: REPRESENTATION OF TRUTH VALUES
There is a truth value object $1 \xrightarrow{t} \Omega$, i.e. there is a one–one correspondence between parts (up to equivalence) of an object X and arrows $X \longrightarrow \Omega$ mediated by pullback along t.

111

Some of the consequences of these axioms have been studied already. Note that *the truth value object* Ω *is not required to be* $1 + 1$. Indeed, in the next section we will see a first example of variable sets, and it will be immediately obvious when we compute the truth value object there that it is *not* $1 + 1$. Thus, we now need to separate two properties that were merged in stating the truth-values axiom in Section 2.4.

Since Ω is a truth value object, the monomapping $0 \longrightarrow \Omega$ has a characteristic mapping called $\neg : \Omega \longrightarrow \Omega$. Precomposing with $1 \overset{t}{\longrightarrow} \Omega$ defines another element of Ω called f for false.

AXIOM: S IS BOOLEAN
Ω *is the following sum:* $\left\{ \begin{smallmatrix} t \\ f \end{smallmatrix} : 1 + 1 \overset{\sim}{\longrightarrow} \Omega \right.$ *where the injections are* t *and* f.

The special toposes which (like S) satisfy this last axiom are called Boolean toposes; they allow the use of classical logic (see Appendix A).

The next axiom has not been explicitly stated until now. We explain why below, but it should be pointed out that there are Boolean toposes that do not satisfy it. For example, the category S/X is always a Boolean topos, but if X has more than one element, then the category does not satisfy this axiom.

AXIOM: S IS TWO-VALUED
Ω *has exactly two elements.*

The two elements of Ω must then be t and f. The axiom is equivalent to the requirement that 1 has exactly two parts.

Exercise 6.1
Prove that $S/2$ is not two-valued. $\Diamond$

AXIOM: THE AXIOM OF CHOICE
Any epimorphism has a section.

Thus, as we have already pointed out, there is a representation of the cells of a partition by a choice of elements of the domain of the partition. This axiom has a very different character from the others. For one thing it can be shown that it implies the Boolean axiom (see McLarty [M92], Theorem 17.9).

The axiom of choice also implies, as a special case, that for any set X the epimorphic part of the unique mapping $X \longrightarrow 1$ is a split epimorphism. Since this image is called the **support** of X, this special property is sometimes referred to by saying that **supports split**.

The principle that nonempty sets have elements is false in most toposes of variable or cohesive sets; moreover, there are several different precise meanings to the term "nonempty" relative to which the principle may be true in various special cases. Rather than just X not equal to 0, the requirement that the terminal map $X \longrightarrow 1$ be an epimorphism is sometimes a more reasonable expression of the idea that X is not empty.

Recall that we assumed in Chapter 1 that S satisfies two other important properties:

(i) 1 is a separator, and
(ii) in S we have $0 \neq 1$.

A topos that satisfies (i) and (ii) is called a **well-pointed** topos. It is a theorem [MM92] that a topos is well-pointed if and only if it is Boolean, two-valued, and supports split. As a result, the two properties we required for S in Chapter 1, namely (i) and (ii), that is that S is well-pointed, are actually consequences of the axioms we have already stated. Conversely, our assumption in Chapter 1 that S is well-pointed implies that S is Boolean and two-valued. It also implies the special case of the axiom of choice called "supports split".

There is one more axiom we will require of S. The axioms so far say nothing about the possibility of mathematical induction. When a starting element of a set and a process for forming new elements of the set (an endo-mapping) are given, a unique sequence determined by the starting element and the process should result. Mathematical practice usually makes the idealization that all such sequences are parameterizations by a single object N; such an N must be "infinite". However, the axioms we have so far do not guarantee the existence of such an object; we will consider it in Section 9.1. In fact, within any category of sets that *has* an infinite object it is possible to find a Boolean topos that *does not* have such an object.

In summary then, we can say precisely what we mean by a **category of abstract sets and arbitrary mappings**. It is a *topos that is two-valued with an infinite object and the axiom of choice* (and hence is also Boolean). Experience shows that mathematical structures of all kinds can be modeled as diagrams in such a topos.

We conclude this section by fulfilling the promises made within Claim 3.4. The first of these is the following:

$$\text{if } X \neq 0, \text{ then } X \text{ has an element } 1 \longrightarrow X.$$

We will show that $X \neq 0 \Rightarrow X \longrightarrow 1$, and then the axiom of choice provides the element we need. Now the (split) image factorization of $X \longrightarrow 1$ as $X \longrightarrow I \longrightarrow 1$ determines a subobject of 1. Since S is two-valued and Boolean, there are only two mappings $1 \longrightarrow 1 + 1 \cong \Omega$. These classify the only two subobjects of 1, which thus must be 1 and 0. Hence, I is either 1 or 0. So we will have completed the proof if we know that $X \longrightarrow 0$ implies $X \cong 0$.

Exercise 6.2

Show that $X \times 0 \xrightarrow{p_1} 0$ is an isomorphism.

Hint: Recall that $Y^0 \cong 1$. Use this to show that any arrow $X \longrightarrow 0$ has an inverse.

$\diamondsuit$

The second promise was that every part $X \xhookrightarrow{i} Y$ has a complement $X' \xhookrightarrow{i'} Y$. One way to see this follows from the next exercise:

Exercise 6.3

Let $A_0 \xrightarrow{i_0} A_0 + A_1 \xleftarrow{i_1} A_1$ be a sum and $B \xrightarrow{g} A_0 + A_1$. In the diagram following with both squares defined to be pullbacks we have that the top row is a sum.

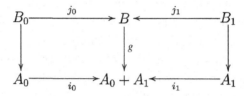

$\diamondsuit$

Now, to fulfill the second promise, we may take the sum diagram for Ω (since $\mathcal{S}$ is Boolean!) as the bottom row and use the characteristic map $Y \xrightarrow{\varphi} \Omega$ for X as g. We conclude that Y is the sum of X and X' since X' is by definition the pullback of φ along the false map f.

6.2 Truth Values for Two-Stage Variable Sets

Perhaps the simplest kind of variable set (beyond the constant kind whose category $\mathcal{S}$ has been studied up to now) is the category $\mathcal{S}^{2^{\text{op}}}$ of two-stage sets with only one connection between the stages. Here we will use the symbols $\mathbf{2} = \boxed{U \longrightarrow 1}$ to suggest a (previous) stage U, which, together with a present stage constitutes a total movement 1. A set X in the category $\mathcal{S}^{2^{\text{op}}}$ of all sets undergoing this movement will be analyzed in terms of constant sets in $\mathcal{S}$ as

Here X_1 in $\mathcal{S}$ is the set of elements of X that persist throughout both stages, X_U in $\mathcal{S}$ is the set of elements of X that persisted through the previous stage, and ξ_X is the map describing the internal structure of X by specifying for each element x of X_1 the element (of X_U) that x "was" during the previous stage U; the result of this specification is denoted by $\xi_X(x)$. A map $X \xrightarrow{f} Y$ in $\mathcal{S}^{2^{\mathrm{op}}}$ is analyzed as a pair f_1, f_U of maps in $\mathcal{S}$ for which $\xi_Y f_1 = f_U \xi_X$, that is the square below commutes:

$$
\begin{array}{ccc}
X & \xrightarrow{\quad f \quad} & Y \\
\\
X_1 & \xrightarrow{\quad f_1 \quad} & Y_1 \\
\xi_X \downarrow & & \downarrow \xi_Y \\
X_U & \xrightarrow{\quad f_U \quad} & Y_U
\end{array}
$$

Exercise 6.4

If $\xi_Y f_1 = f_U \xi_X$ and $\xi_Z g_1 = g_U \xi_Y$, then

$$\xi_Z(g_1 f_1) = (g_U f_U)\xi_X$$

$\Diamond$

By the exercise, we get a well-defined operation of composition of maps in $\mathcal{S}^{2^{\mathrm{op}}}$:

$$(gf)_1 = g_1 f_1 \qquad (gf)_U = g_U f_U$$

The terminal set 1 of $\mathcal{S}^{2^{\mathrm{op}}}$ is (in its $\mathcal{S}$-analysis)

$$
1 = \begin{array}{c} 1 \\ \downarrow {\scriptstyle 1_1} \\ 1 \end{array}
$$

since, for any X, there is a unique commutative square

$$
\begin{array}{ccc}
X_1 & \longrightarrow & 1 \\
\xi_X \downarrow & & \downarrow \\
X_U & \longrightarrow & 1
\end{array}
$$

in $\mathcal{S}$. To what extent do the maps

$$1 \xrightarrow{x} X$$

in $\mathcal{S}^{2^{\mathrm{op}}}$ represent the elements of X? The equation required of any map in this category reduces to

$$
\begin{array}{ccc}
1 & \xrightarrow{\quad x_1 \quad} & X_1 \\
\downarrow & & \downarrow \xi_X \\
1 & \xrightarrow{\quad x_U \quad} & X_U
\end{array}
$$

in this case; thus, x is entirely determined by x_1, and conversely any element x_1 of X_1 in $\mathcal{S}$ (no condition) determines, using ξ_X, a unique x_U making the square commutative, i.e. determines a unique $1 \xrightarrow{x} X$ in $\mathcal{S}^{2^{\mathrm{op}}}$. Thus, we may briefly say that the maps $1 \xrightarrow{x} X$ "are" the elements of X that live through both stages.

How can we represent the elements of X in their aspect of existence only through the previous stage U? To that end consider the variable set

$$U = \begin{matrix} 0 \\ \downarrow \\ 1 \end{matrix}$$

which has no elements that persist through both the present and previous stages but has one element throughout the previous stage. This U then is a *nonzero* variable set with *no* element in the narrow sense. The property that nonzero abstract sets do have elements, as discussed in the previous section, is seen to be already violated with this simple variation. Now for any X, the maps $U \xrightarrow{x} X$ in $\mathcal{S}^{2^{\mathrm{op}}}$ may be analyzed in $\mathcal{S}$ as

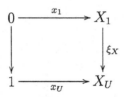

but x_1 is unique, and the square commutes no matter what element x_U of X_U is chosen. Thus, we may say the maps $U \xrightarrow{x} X$ "are" the elements of X that existed at the previous stage.

Exercise 6.5
There is exactly one map $U \xrightarrow{\xi} 1$ in $\mathcal{S}^{2^{\mathrm{op}}}$. Moreover, given any $1 \xrightarrow{x} X$, the composition $U \xrightarrow{\xi} 1 \xrightarrow{x} X$ in $\mathcal{S}^{2^{\mathrm{op}}}$ represents the x_U that the ξ_X specifies as the element that x "was" throughout U.

(Draw the appropriate diagram to verify the statement.)   ◊

Exercise 6.6
A map $A \xrightarrow{i} X$ in $\mathcal{S}^{2^{\mathrm{op}}}$ is a part of X if and only if the following three conditions are all satisfied.

$$A_1 \text{ is a part of } X_1 \text{ in } \mathcal{S}$$
$$A_U \text{ is a part of } X_U \text{ in } \mathcal{S}$$
$$\xi_X i_1 = i_U \xi_A$$

◊

Moreover, ξ_A is uniquely determined (if it exists) by X, i_1, i_U; the condition that ξ_A exists is

$$\forall x \in X_1[x \in A_1 \Longrightarrow \xi_X x \in A_U]$$

$$\begin{array}{ccc} A_1 & \xrightarrow{\ i_1\ } & X_1 \\ \xi_A \downarrow & & \downarrow \xi_X \\ A_U & \xrightarrow[\ i_U\ ]{} & X_U \end{array}$$

Now we will calculate the truth-value variable set $\mathcal{P}(1) = \mathcal{P}_2(1)$ in $\mathcal{S}^{2^{\mathrm{op}}}$ that precisely represents all parts in that category through characteristic functions

$$\frac{X \longrightarrow \mathcal{P}(1)}{? \ \hookrightarrow\ X} \quad \downarrow\uparrow$$

An arbitrary part $A \xrightarrow{i_A} X$ may involve elements of X that are not now in A but were previously in A. Yet the characteristic function φ of A must be defined for all elements of X, and of course φ must be a map in $\mathcal{S}^{2^{\mathrm{op}}}$. By taking $X = 1$ and $X = U$ above, this forces

$$\mathcal{P}(1) = \boxed{0\ U\ 1}\ \begin{array}{c}\\ \downarrow\downarrow\downarrow \end{array}\ \boxed{0 \quad 1}$$

to have three truth values globally, but only two truth values in the previous stage. For any $A \hookrightarrow X$, one can define φ at the "present" stage by

$$\varphi_1 x = \begin{cases} 0 \text{ if } x \notin A,\ \xi_X(x) \notin A \\ U \text{ if } x \notin A,\ \xi_X(x) \in A \\ 1 \text{ if } x \in A \end{cases}$$

Exercise 6.7
Define also φ_U and show that $\varphi_U \xi_X = \xi_{\mathcal{P}(1)} \varphi_1$. For each of the two types of elements of X,

$$x \in A \Longleftrightarrow \varphi x = 1$$

$\Diamond$

6.3 Additional Exercises

Exercise 6.8
The categories $\mathcal{S}/X$ satisfy most of the axioms in Section 6.1, as we have seen in Exercises 3.48, 4.46, 4.52, 5.12. Show that $\mathcal{S}/X$ has a truth-value object, namely,

the projection $2 \times X \longrightarrow X$. Thus, S/X fails to be a category of abstract sets and mappings only in that it is not two-valued.

Exercise 6.9

(a) Show that there is a (*diagonal*) functor (see Appendix C.1) denoted Δ_X from S to S/X whose value at a set A is the object of S/X given by the projection mapping $A \times X \longrightarrow X$. (So define Δ_X also on mappings and show that it satisfies the equations.)

(b) Show that there is a functor Σ_X (for *sum*, why?) from S/X to S that sends an object $Y \longrightarrow X$ of S/X to the set Y.

These two functors have the following important relationship:

(c) Show that for any set A in S and object $Y \xrightarrow{\varphi} X$ in S/X there is a one–one correspondence between mappings in S from $\Sigma_X(\varphi)$ to A and mappings in S/X from φ to $\Delta_X(A)$. (Σ_X is *left adjoint* – see Appendix C.1 – to Δ_X.)

Exercise 6.10
Show that $S^{2^{\mathrm{op}}}$ has all finite limits and colimits.

Hint: They are computed "pointwise".

Exercise 6.11
The previous exercise and the description in Section 6.2 of $\mathcal{P}(1)$ show that $S^{2^{\mathrm{op}}}$ is not Boolean. Show that it is not two-valued either.

Exercise 6.12

(a) Show that epimorphisms in $S^{2^{\mathrm{op}}}$ are pointwise. By "pointwise" here we mean a property holding at both 1 and U. Thus, epimorphisms are pointwise means that $X \xrightarrow{f} Y$ is epi in $S^{2^{\mathrm{op}}}$ iff f_1 and f_U are epi in S.

(b) Show that supports split in $S^{2^{\mathrm{op}}}$, *but*

(c) Show that the axiom of choice fails in $S^{2^{\mathrm{op}}}$.

   Hint: There is a nonsplit epimorphism between objects with no more than two elements at each vertex.

Exercise 6.13
Show how to construct mapping sets in $S^{2^{\mathrm{op}}}$.

Hint: $(Y^X)_U = Y_U^{X_U}$ and $(Y^X)_1$ is a certain set of pairs of mappings.

Exercise 6.14

(a) Show that there is a (*diagonal*) functor (see Appendix C.1) denoted Δ from $\mathcal{S}$ to $\mathcal{S}^{\mathbf{2}^{op}}$ whose value at a set A is the object of $\mathcal{S}^{\mathbf{2}^{op}}$ given by the identity mapping $A \longrightarrow A$. (So define Δ also on mappings and show it satisfies the equations.)

(b) Show that there is a functor **dom** from $\mathcal{S}^{\mathbf{2}^{op}}$ to $\mathcal{S}$ that sends an object $X = X_1 \xrightarrow{\xi_X} X_U$ of $\mathcal{S}^{\mathbf{2}^{op}}$ to the set X_1.

(c) Show that there is a functor **cod** from $\mathcal{S}^{\mathbf{2}^{op}}$ to $\mathcal{S}$ that sends an object $X = X_1 \xrightarrow{\xi_X} X_U$ of $\mathcal{S}^{\mathbf{2}^{op}}$ to the set X_U.

These three functors have the following important relationships. Show that for any set A in $\mathcal{S}$ and object $X = X_1 \xrightarrow{\xi_X} X_U$ of $\mathcal{S}^{\mathbf{2}^{op}}$ there are one–one correspondences

(i) between mappings in $\mathcal{S}$ from $\mathbf{cod}(X)$ to A and mappings in $\mathcal{S}^{\mathbf{2}^{op}}$ from X to $\Delta(A)$ (**cod** is *left adjoint* – see Appendix C.1 – to Δ) and

(ii) between mappings in $\mathcal{S}^{\mathbf{2}^{op}}$ from $\Delta(A)$ to X and mappings in $\mathcal{S}$ from A to $\mathbf{dom}(X)$ (**dom** is *right adjoint* to Δ).

Exercise 6.15

If M is a monoid, the category of M-sets has finite limits and colimits as well as mapping sets (Exercises 3.51, 4.47, 5.13), but does not satisfy the axiom of choice (4.54). It does have a truth-value object. First, note that a subobject of an M-set X is simply a part $A \longhookrightarrow X$ of X that is closed under the right action of M. For its characteristic function φ_A we send $x \in X$ to the set I of $i \in M$ such that $xi \in A$. This I is a part of M closed under right multiplication by all of M (called a "right ideal"). The set of right ideals of M, denoted Ω_M, has a right action by M given by $Im = \{j \in M \mid mj \in I\}$, and so it is an M-set. It is the object we seek; M itself is an element of Ω_M that plays the role of "true".

Show that, with the characteristic mappings just outlined Ω_M is indeed the truth-value object.

Thus, the category of M-sets is a topos.

7

Consequences and Uses of Exponentials

7.1 Concrete Duality: The Behavior of Monics and Epics under the Contravariant Functoriality of Exponentiation

Any conceivable cancellation law states in effect that some algebraic process or other is injective. This is most evident in the definition of the concept of monomorphism, whereby the "injectivity" of the algebraic process of composing with f on the left turns out to be equivalent to injectivity of f itself as a map. In terms of function spaces we can express the equivalence

$$f \text{ injective} \Longleftrightarrow f \text{ monomorphic}$$

simply by

$$X \xrightarrow[\text{injective}]{f} Y \Longleftrightarrow \forall T \left[X^T \xrightarrow[\text{injective}]{f^T} Y^T \right]$$

since we see that to say that f^T is injective in its action on elements $1 \longrightarrow X^T$ defined on 1 is equivalent to saying that for all x_1, x_2

$$T \underset{x_2}{\overset{x_1}{\rightrightarrows}} X \xrightarrow{f} Y, \quad fx_1 = fx_2 \Longrightarrow x_1 = x_2$$

if we only recall that the action of f^T is

$$f^T \ulcorner x \urcorner = \ulcorner fx \urcorner \text{ for all } T \xrightarrow{x} X$$

But what if f is epimorphic? Since the statement

$$\varphi_1 f = \varphi_2 f \Longrightarrow \varphi_1 = \varphi_2$$

is also a cancellation property, it also expresses that some process is *injective*, and if we look a little more closely, we see that the process in question is the one represented by the *contravariant* functoriality of mapping sets. For later use it is preferable to state this fact as property relative to a given object V.

Proposition 7.1: *For a given* $X \xrightarrow{f} Y$ *and a given* V,

$$V^Y \xrightarrow{V^f} V^X$$

is injective (on elements) exactly when

$$\varphi_1 f = \varphi_2 f \implies \varphi_1 = \varphi_2$$

holds for any $Y \underset{\varphi_2}{\overset{\varphi_1}{\rightrightarrows}} V$.

Proof:

$$V^f \ulcorner \varphi \urcorner = \ulcorner \varphi f \urcorner$$

∎

Corollary 7.2: $X \xrightarrow{f} Y$ *is epimorphic if and only if*

$$V^Y \xrightarrow{V^f} V^X$$

is injective for all V.

It will be recalled that the definition of (for example) "epimorphism" is the *formal dual* of the definition of "monomorphism" in the sense that one simply reverses all arrows in the relevant diagrams; of course if the original diagrams had been given *specific* interpretation in terms of *specific* sets and mappings, such interpretation is lost when we pass to this formal dual in that the formal dualization process in itself does *not* determine specific sets and specific mappings that interpret the dualized statement. On the other hand, for any given V, we do have the process of contravariant functoriality that for any specific diagram, say

$$X \rightrightarrows Y$$

produces a specific diagram (with "bigger" sets!) in which all arrows have been reversed

$$V^X \leftleftarrows V^Y$$

and that satisfies (at least) all the commutativities (= statements about equality of compositions) satisfied by the original diagram except, of course, that the *order* of the composition has been reversed. This is often referred to as "concrete duality with respect to V" or "dualizing into V". Not every statement will be taken into its formal dual by the process of dualizing with respect to V, and indeed a large part of the study of mathematics

$$\text{s p a c e} \quad \text{vs.} \quad \text{q u a n t i t y}$$

and of logic

$$\text{t h e o r y} \quad \text{vs.} \quad \text{e x a m p l e}$$

may be considered as the detailed study of the extent to which formal duality and concrete duality into a favorite V correspond or fail to correspond.

In the context of constant sets, the choice $V = 2$ is the starting point of many such duality theories.

Theorem 7.3: $X \xrightarrow{f} Y$ *is epimorphic if and only if* $2^Y \xrightarrow{2^f} 2^X$ *is monomorphic.*

Proof: The "only if" direction is a special case of (the internalization of) the definition of "epic".

Conversely, if $2^Y \xrightarrow{2^f} 2^X$ is monomorphic, then f satisfies the (restricted to 2) right-cancellation property

$$X \xrightarrow{f} Y \underset{\varphi_2}{\overset{\varphi_1}{\rightrightarrows}} 2, \quad \varphi_1 f = \varphi_2 f \implies \varphi_1 = \varphi_2$$

But we can take

$$\varphi_1 = \text{characteristic function of the image of } f$$
$$\varphi_2 = \text{identically true,}$$

which will surely satisfy $\varphi_1 f = \varphi_2 f$; if we apply the assumed cancellation property, we see that f is epimorphic (hence intuitively that the image of f fills up the whole of Y). ∎

Because the dualization functor is not reversible without modification, the following further duality property does not follow from the previous theorems.

Theorem 7.4: $X \xrightarrow{f} Y$ *is a monomorphism if and only if* $2^Y \xrightarrow{2^f} 2^X$ *is epimorphic.*

Proof: If $T \overset{x_1}{\underset{x_2}{\rightrightarrows}} X \xrightarrow{f} Y$ have equal composites, then by functoriality so do

$$2^Y \longrightarrow 2^X \rightrightarrows 2^T$$

and thus if 2^f is epimorphic, then $2^{x_1} = 2^{x_2}$, from which we will conclude shortly that $x_1 = x_2$, so that f is monomorphic.

Conversely, if f is assumed monomorphic, we need to conclude that

$$2^Y \xrightarrow{2^f} 2^X$$

is epimorphic, for which it suffices to show that 2^f is surjective. So assume $\ulcorner\varphi\urcorner$ is any element of 2^X; we must show that there is at least one $\ulcorner\psi\urcorner$ for which $2^f\ulcorner\psi\urcorner = \ulcorner\varphi\urcorner$, i.e. for which

But φ is the characteristic function of a part i of X, and by assumption f is a part of Y; thus, the composite fi is a part of Y; we naturally guess that taking ψ to be the characteristic function of this composite part will satisfy our requirement, which we can prove as follows:

$$\varphi x = \text{true} \Longleftrightarrow x \in i$$
$$\psi y = \text{true} \Longleftrightarrow y \in fi$$

$$S \overset{i}{\hookrightarrow} X \overset{f}{\hookrightarrow} Y$$

Calculate, for any x

$$(\psi f)x = \text{true}$$
$$\Updownarrow$$
$$\psi(fx) = \text{true}$$
$$\Updownarrow$$
$$fx \in fi$$
$$\Updownarrow$$
$$x \in i$$
$$\Updownarrow$$
$$\varphi x = \text{true}$$

Hence, $\psi f = \varphi$ as required.

We return to showing that

$$2^{x_1} = 2^{x_2} \Rightarrow x_1 = x_2$$

A useful tool is the **singleton map** $\{\ \}$ defined as the exponential transpose of the characteristic function of the diagonal

$$\frac{\dfrac{X \xrightarrow{\ \delta_X\ } X \times X}{X \times X \longrightarrow 2}}{X \xrightarrow{\ \{\ \}\ } 2^X}$$

In other words if $\{x\} = \ulcorner \varphi \urcorner$ and i is the part of X with characteristic function φ, then for any x'

$$x' \in i \iff x' = x$$

Now if we assume that $T \underset{x_2}{\overset{x_1}{\rightrightarrows}} X$ are such that in $2^X \rightrightarrows 2^T$ we have $2^{x_1} = 2^{x_2}$, then composing with $X \xrightarrow{\{\ \}} 2^X$ we get equal maps $X \rightrightarrows 2^T$ and hence equal maps $T \times X \rightrightarrows 2$ and therefore equal parts of $T \times X$. But it is easily seen that the parts of $T \times X$ arising in this way are actually the graphs of the corresponding maps $T \rightrightarrows X$; if the graphs are equal, then the corresponding maps are equal, as was to be shown. ∎

Theorem 7.4 about 2 in the category of constant sets becomes a definition of a special kind of object V in more general categories.

Definition 7.5: *An object V is an* **injective object** *if for every monomorphism* $X \xrightarrow{\ f\ } Y$ *and every* $X \xrightarrow{\ \varphi\ } V$ *there exists* $Y \xrightarrow{\ \overline{\varphi}\ } V$ *such that*

$$\overline{\varphi} f = \varphi$$

In case the category has exponentiation, V is injective if and only if for every monomorphism $X \xrightarrow{f} Y$, $V^Y \xrightarrow{V^f} V^X$ is surjective (on elements $1 \longrightarrow V^X$). It is not surprising that most of the concrete dualities in mathematics and logic that work well involve dualizing into an injective object V.

Another twist on the same underlying problem is the following: Let V be any given object and define $X \xrightarrow{\ f\ } Y$ to be **co-surjective** relative to V to mean

$$\forall \varphi \, \exists \overline{\varphi} [\overline{\varphi} f = \varphi]$$

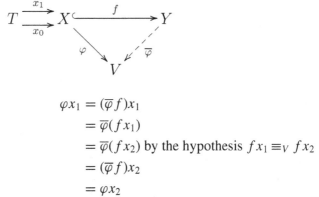

By abstract duality plus justifiable prejudice we might expect co-surjectivity to be similar to injectivity for a map. To make this precise, consider one more definition:

For $T \underset{x_2}{\overset{x_1}{\rightrightarrows}} X$, say that $x_1 \equiv_V x_2$ if and only if

$$\forall X \xrightarrow{\ \varphi\ } V [\varphi x_1 = \varphi x_2]$$

This $\equiv_V$ may be read as "congruent modulo V" or more simply as "equal insofar as V-valued properties can distinguish".

Proposition 7.6: *If f is co-surjective relative to V, then f is "monic modulo V," i.e.,*

$$f x_1 \equiv_V f x_2 \Longrightarrow x_1 \equiv_V x_2$$

Proof: The hypothesis of the implication refers of course to testing relative to all $Y \longrightarrow V$. We must show $\varphi x_1 = \varphi x_2$ for any given $X \xrightarrow{\ \varphi\ } V$. But by co-surjectivity φ can be extended to a $\overline{\varphi}$

$$T \underset{x_0}{\overset{x_1}{\rightrightarrows}} X \overset{\ f\ }{\hookrightarrow} Y$$

$$
\begin{aligned}
\varphi x_1 &= (\overline{\varphi} f) x_1 \\
&= \overline{\varphi}(f x_1) \\
&= \overline{\varphi}(f x_2) \text{ by the hypothesis } f x_1 \equiv_V f x_2 \\
&= (\overline{\varphi} f) x_2 \\
&= \varphi x_2
\end{aligned}
$$
■

Without yet going into detail concerning the modifications necessary to reverse the concrete duality, we note that double dualization $V^{(V^X)}$ depends only on the dual V^X and is a *covariant* functor in that $X \xrightarrow{f} Y$ induces

$$V^{V^X} \longrightarrow V^{V^Y}$$

in the same direction. Moreover, there is for each X a map

$$X \xrightarrow{\;(\hat{\,})\;} V^{V^X}$$

defined by

$$x \longmapsto \ulcorner \varphi \longmapsto \varphi x \urcorner$$

and often written (ignoring the distinction between a map and the corresponding element of a mapping set) as $\hat{x}(\varphi) = \varphi(x)$. The map $(\hat{\,})$, sometimes called the Fourier transform or the Dirac delta, is **natural** in the sense that for any map $X \xrightarrow{f} Y$ the diagram

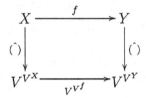

is commutative. In case V is a coseparator, the map $(\hat{\,})$ is monic and thus X is (the domain of) a *part* of the double dual of X. The fact that this double dual is much larger than X can often be overcome, and X can actually be recovered from the knowledge of its dual $A = V^X$ by noting that the "variable quantities" A and the "constant quantities" V have some algebraic structure in common and considering only the part $A^* = \mathrm{Hom}\,(A, V) \hookrightarrow V^A$ whose members correspond to those maps $A \longrightarrow V$ that preserve this algebraic structure; the thus-tempered $X \xrightarrow{(\hat{\,})} (V^X)^*$ has a better chance of being invertible, as we shall see in Section 8.4.

7.2 The Distributive Law

The distributive law states that a natural map

$$A \times X_1 + A \times X_2 \longrightarrow A \times (X_1 + X_2)$$

is actually invertible. Here the $\times$ and $+$ denote product and coproduct ($=$ sum in the case of sets), and the natural map exists whenever product and coproduct exist. In fact the natural map in question comes from a special case of the following observation: to define a map $? \longrightarrow A \times S$ whose *codomain* is a product is equivalent to defining maps from $?$ to each of the factors, whereas to define a map $P_1 + P_2 \longrightarrow ??$ whose *domain* is a coproduct is equivalent to defining maps from each of the summands to $??$; hence, in case we have both that the codomain is known to be a product *and* that the domain is known to be a coproduct, then defining a map is equivalent to defining a "matrix" of (smaller) maps specifying each component of the value of the big map for each kind (summand) of input. For example, any map

$$P_1 + P_2 \longrightarrow A \times S$$

(between the indicated combinations of any four given sets) is specified by a 2×2 matrix whose entries would be the four possible threefold composites with projections and injections.

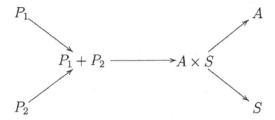

In the case at hand, where $S = X_1 + X_2$, $P_k = A \times X_k$, we have enough structure to actually specify such a matrix, where

$$P_1 \longrightarrow A \text{ is } A \times X_1 \xrightarrow{\text{proj}} A$$
$$P_2 \longrightarrow A \text{ is } A \times X_2 \xrightarrow{\text{proj}} A$$
$$P_1 \longrightarrow S \text{ is } A \times X_1 \xrightarrow{\text{proj}} X_1 \xrightarrow{\text{inj}} X_1 + X_2$$
$$P_2 \longrightarrow S \text{ is } A \times X_2 \xrightarrow{\text{proj}} X_2 \xrightarrow{\text{inj}} X_1 + X_2$$

Of course, the natural map just described in detail is the one corresponding to the intuitive picture of a product as a rectangular arrangement and of a coproduct of sets as a disjoint union:

However, although the "rectangular arrangement" picture of the elements of a product is justified by the defining property of products, the "disjoint sum" picture of the elements of a coproduct is NOT justified by the defining property *in itself* of coproducts. This is because the defining property of a coproduct $X_1 + X_2 = S$ refers to co-elements of S,

$$S \longrightarrow V$$

and says nothing in itself about how to determine the elements

$$T \longrightarrow S$$

that S might have. Indeed, in other important categories the coproduct may be quite unlike a disjoint sum; for example, in the basic category of linear algebra namely, the category of vector spaces and linear transformations (also called linear spaces and linear maps) the coproduct is the same space as the product (see Exercise 3.46), and hence neither the distributive law nor "disjointness" could hold. Thus, the invertibility of the natural map $A \times X_1 + A \times X_2 \longrightarrow A \times (X_1 + X_2)$ in the context of nonlinear spaces and nonlinear maps (in particular for abstract sets and arbitrary maps) must be due (like the disjointness of $+$) to some further feature of these categories beyond the mere existence of product and coproduct in themselves. The existence of *exponentiation* turns out to supply this feature. For to show that the natural map has an inverse we must construct a candidate

$$A \times (X_1 + X_2) \longrightarrow A \times X_1 + A \times X_2$$

and then verify that the candidate really is inverse; the construction is the step that requires some additional ingredient. But note that the required candidate should be a map whose *domain is a product* and that since the crucial feature of the exponentiation axiom is an invertible process, we could just as well view that axiom as a prescription for constructing lower-order maps $A \times S \longrightarrow Y$ by instead constructing higher-order maps $S \longrightarrow Y^A$. This apparently perverse interpretation of the exponentiation axiom turns out to be exactly what is needed in this case (and in many other cases) since it enables us to gain access to certain special structures that S has. That is, the proof of the distributive law can be achieved by constructing an inverse map

$$A \times (X_1 + X_2) \xrightarrow{\ ?\ } A \times X_1 + A \times X_2$$

as we have seen. But by the exponentiation axiom, we can equivalently construct

$$X_1 + X_2 \xrightarrow{\ ?\ } (A \times X_1 + A \times X_2)^A$$

which (by the special coproduct structure of $S = X_1 + X_2$) is equivalent to the problem of constructing *two* maps (for $k = 1, 2$)

$$X_k \xrightarrow{\ ?\ } (A \times X_1 + A \times X_2)^A$$

By applying exponentiation again, but in the opposite direction, this is equivalent to constructing two maps

$$A \times X_k \longrightarrow A \times X_1 + A \times X_2$$

But such maps are staring us in the face – the two injections! Hence, we can construct the required candidate map

$$A \times (X_1 + X_2) \longrightarrow A \times X_1 + A \times X_2$$

which can be shown to behave as expected,

$$\langle a, i_k x \rangle \mapsto i_k \langle a, x \rangle$$

on elements (when $1 \xrightarrow{x} X_k$).

Exercise 7.7
Verify that the candidate map just constructed really is the two-sided inverse of the natural map

$$A \times X_1 + A \times X_2 \longrightarrow A \times (X_1 + X_2)$$

$\Diamond$

Exercise 7.8
Prove (using exponentiation) that

$$0 \xrightarrow{\sim} A \times 0$$

$\Diamond$

Exercise 7.9
Prove (using exponentiation) that if $X \xrightarrow{f} Y$ is an epimorphic map, then

$$A \times X \xrightarrow{1_A \times f} A \times Y$$

is also epimorphic.

$\Diamond$

7.3 Cantor's Diagonal Argument

Over a century ago Georg Cantor proved an important theorem that includes the result

$$X < 2^X$$

for all sets X (see Def. 7.15 below). This result, well-known for finite sets X, was quite new for infinite sets since it showed that some infinities are definitely bigger than others (even though many constructions on infinite sets X tend to give sets of the same cardinality, that is,

$$2 \times X \cong X, \qquad X^2 \cong X$$

hold for infinite abstract sets). Indeed, this specific and fundamental construction leads to a potentially infinite sequence of larger and larger infinities

$$X < 2^X < 2^{2^X} < 2^{2^{2^X}} < \dots$$

An even more fundamental consequence of Cantor's theorem is the obvious conclusion that there *cannot exist a "universal set" V* for which *every* set X appears as the domain of a part of V

$$X \lhook\joinrel\longrightarrow V$$

because if there were such a V, we could take $X = 2^V$ to reach a contradiction since (see Theorem 7.4) the restriction map

$$2^V \longrightarrow 2^X$$

is surjective, but Cantor showed that no map $X \longrightarrow 2^X$ is surjective.

Cantor's method for proving this theorem is often called the "diagonal argument" even though the diagonal map δ_X is only one of two equally necessary pillars on which the argument stands, the second being a fixed-point-free self-map τ (such as logical negation in the case of the set 2). This diagonal argument has been traced (by philosophers) back to ancient philosophers who used something like it to mystify people with the Liar's paradox. Cantor, however, used his method to prove positive results, namely inequalities between cardinalities. The philosopher Bertrand Russell, who was familiar with Cantor's theorem, applied it to demonstrate the inconsistency of a system of logic proposed by the philosopher Frege; since then philosophers have referred to Cantor's theorem as Russell's paradox and have even used their relapse into the ancient paradox habit as a reason for their otherwise unfounded rumor that Cantor's set theory might be inconsistent. (Combatting this rumor became one of the main preoccupations of the developers of the *axiomatized* set theories of Zermelo, Fraenkel, von Neumann, and Bernays [Sup72]. This preoccupation assumed such an importance that the use of such axiom systems for *clarifying* the role of abstract sets as a *guide* to mathematical subjects such as geometry, analysis, combinatorial topology, etc., fell into neglect for many years.)

Around 1930 both Gödel and Tarski again used exactly the same diagonal argument of Cantor, except in categories of a more linguistic nature than the category of sets, to obtain their famous results to the effect that (Gödel) for any proposed axiom system for number theory there will always be further truths about the number system that do not follow as theorems in that axiom system, and (Tarski) even though it is easy to construe mathematical statements φ as mathematical entities $\ulcorner \varphi \urcorner$, there can be no definition of a single "truth predicate" T such that, for any statement φ and every specific mathematical entity x,

$$\ulcorner\varphi\urcorner Tx \Longleftrightarrow \varphi(x)$$

is a mathematical theorem.

We will first prove the theorem in a still more positive form as a fixed point theorem, of which Cantor's theorem will be the contrapositive. Cantor himself proved cardinality inequalities not only for the set 2^X of subsets but also for the set $\mathbb{R}^X$ of real-valued functions, and correspondingly our fixed-point theorem will deal with objects Y more general than $Y = 2$.

Definition 7.10: *A self-map $Y \xrightarrow{\tau} Y$ of an object Y is said to have $1 \xrightarrow{y} Y$ as a* **fixed point** *if and only if $\tau y = y$. Thus, τ is said to be* **fixed-point-free** *if and only if $\forall 1 \xrightarrow{y} Y[\tau y \neq y]$. At the other extreme, an object Y is said to have the* **fixed-point property** *if and only if every self-map τ has at least one fixed-point y.*

Remark 7.11: Although the fixed-point-property is so rare as to be uninteresting for the category of abstract constant sets and arbitrary maps, it is much more frequent and useful in the category of continuous spaces and maps, where Brouwer proved that the n-dimensional ball has the fixed point property; for $n = 1$, this fact implies existential statements such as Rolle's theorem, which involves continuous maps of the interval $[-1, 1]$ into itself.

Theorem 7.12: *Suppose there is an X and a map φ*

$$X \times X \xrightarrow{\varphi} Y$$

such that for every $X \xrightarrow{f} Y$ there is at least one $1 \xrightarrow{a} X$ such that

$$f = \varphi(a, -)$$

Then Y has the fixed-point property.

Proof: Consider any $Y \xrightarrow{\tau} Y$. We must show that τ has a fixed point. Define an f by the triple composite

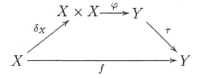

In other words, for all x

$$fx = \tau\varphi(x, x)$$

Now by the assumed property of φ, this f must be "φ-represented" by some a:

$$fx = \varphi(a, x)$$

for all x. Hence,

$$\tau\varphi(x, x) = \varphi(a, x)$$

for all x. In particular, if we take $x = a$,

$$\tau\varphi(a, a) = \varphi(a, a)$$

which means that if we define y in terms of a by commutativity of

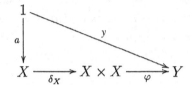

then y is a fixed point of τ

$$\tau y = y$$

∎

Corollary 7.13: *(Cantor) If Y has at least one self-map τ with no fixed points, then for every X and for every*

$$X \xrightarrow{\Phi} Y^X$$

Φ is not surjective.

Proof: Surjectivity of Φ expressed by the diagram

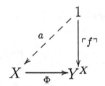

is exactly the property assumed in the theorem for the corresponding φ for which $\Phi = \ulcorner\varphi\urcorner$. The corollary is the contrapositive of the theorem, and any statement implies its contrapositive.   ∎

Corollary 7.14: *There is no surjective map*

$$X \longrightarrow 2^X$$

nor is there any surjective map

$$X \longrightarrow \mathbb{R}^X$$

where $\mathbb{R}$ is the set of real numbers.

Proof: Logical negation $2 \xrightarrow{\tau} 2$ is fixed-point-free. Indeed, $\tau(\text{false}) \neq \text{false}$, and $\tau(\text{true}) \neq (\text{true})$. For the second statement, all we need to know about $\mathbb{R}$ is that it has a self-map τ such as

$$\tau(x) = x + 1$$

for which $\tau(x) \neq x$ for all $1 \xrightarrow{x} \mathbb{R}$. ∎

Definition 7.15: *For sets X and Y we write $X \leq Y$ when there exists at least one monomapping from X to Y. We write $X < Y$ to mean $X \leq Y$ and that moreover no surjective maps $X \longrightarrow Y$ exist.*

Corollary 7.16:

$$X < 2^X \text{ for all } X$$

Indeed

$$X < Y^X \text{ for any } Y \text{ with } 2 \leq Y$$

∎

Another frequently cited application of Cantor's argument shows that there are strictly more real numbers than rational numbers. This can be established in three steps as follows:

$$\mathbb{Q} \cong \mathbb{N} < \{0, 1, 2, 3, 4, 5, 6, 7, 8\}^{\mathbb{N}} \hookrightarrow \mathbb{R}$$

That is, we separately establish (by a snakelike counting of fractions) that the set of rational numbers is isomorphic as an abstract set with the set of natural numbers, and we note that among the reals there are those (nonnegative ones) whose decimal expansion involves no 9's. The latter set is equivalent to the function space Y^X indicated with a sequence $\mathbb{N} \xrightarrow{a} \{0, 1, \ldots, 8\}$ mapping to the real with decimal expansion

$$a_0 \cdot a_1 a_2 a_3 \ldots = \sum_{n=0}^{\infty} a_n 10^{-n}$$

Thus, we can apply our argument above with $X = \mathbb{N}$, $Y = \{0, \ldots, 8\}$, noting that the finite set Y does indeed have endomaps that move every element.

7.4 Additional Exercises

Exercise 7.17
Show that the truth-value object Ω is an injective object in any topos. More generally show that the mapping set Ω^X is an injective object for any object X.

Exercise 7.18
Find all of the injective objects in the categories $\mathcal{S}/X$.

Hint: Start with the case $X = 1$ (i.e., the category of abstracts sets and mappings).

Exercise 7.19
Find all of the injective objects in the category $\mathcal{S}^{2^{\mathrm{op}}}$.

Exercise 7.20
The distributive law holds in any topos. Verify this explicitly for $\mathcal{S}^{2^{\mathrm{op}}}$.

Exercise 7.21
Categories with finite sums and products (including 0 and 1!) in which the distributive law holds are extremely important in theoretical computer science. For example (see Walters [Wal91]), let $f, g : A \longrightarrow B$ and suppose $\varphi : A \longrightarrow 2$ is a "test function". Show that the triple composite

$$A \xrightarrow{\langle 1, \varphi \rangle} A \times 2 \xrightarrow{\sim} A + A \xrightarrow{\left\{ \begin{smallmatrix} f \\ g \end{smallmatrix} \right.} B$$

can be interpreted as "if φ then f else g". Indeed, if we call the composite $h : A \longrightarrow B$, then

$$h(a) = \begin{cases} f(a) & \text{if } \varphi(a) = \text{true} \\ g(a) & \text{otherwise} \end{cases}$$

Exercise 7.22

The concept of **natural transformation** τ from a functor F to a functor G is defined in Appendix C.1.

(a) Show that there is a natural transformation α from the functor **dom**: $\mathcal{S}^{2^{op}} \longrightarrow \mathcal{S}$ to the functor **cod** : $\mathcal{S}^{2^{op}} \longrightarrow \mathcal{S}$ (see Exercise 6.14) whose component for an object X of $\mathcal{S}^{2^{op}}$ is the mapping $X_1 \xrightarrow{\xi_X} X_U$.

(b) Show that for any set B there is a natural transformation β from the functor $- \times B : \mathcal{S} \longrightarrow \mathcal{S}$ (see Exercise 5.17) to the *identity functor* $1_{\mathcal{S}} : \mathcal{S} \longrightarrow \mathcal{S}$.

Hint: The components are projections.

(c) Functors can be composed. For example, the composite of the functors $\Delta : \mathcal{S} \longrightarrow \mathcal{S} \times \mathcal{S}$ and $- \times - : \mathcal{S} \times \mathcal{S} \longrightarrow \mathcal{S}$ (see Exercise 5.17) is the functor $(- \times -)\Delta : \mathcal{S} \longrightarrow \mathcal{S}$ whose value at a set X is $X \times X$. Show that the product projections provide the components of *two* natural transformations from the functor $(- \times -)\Delta$ to the identity functor $1_{\mathcal{S}} : \mathcal{S} \longrightarrow \mathcal{S}$.

8

More on Power Sets

8.1 Images

We have been introduced to the contravariant functoriality of 2^X, which should be understood *both* as

(1) a special case of the composition-induced contravariant functoriality ($X \longrightarrow Y$ induces $V^Y \longrightarrow V^X$) of V-valued function spaces (5.5), and
(2) the operation of inverse image on parts

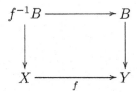

internalized using the special property that $V = 2$ has of encoding parts via characteristic functions (2.30, 2.34).

The covariant functor 2^{2^X} (obtained by composing the contravariant $2^{(\ )}$ with itself) can be interpreted to consist of all "classes" of parts of X, since a map $2^X \xrightarrow{\alpha} 2$ is the characteristic function of a definite part (or class) of the set 2^X of parts of X.

But we now want to consider an important way of making the smaller 2^X a *covariant* functor of X.

First let us agree to use a somewhat more convenient notation for parts. A part of X consists of two components:

(1) the underlying set $|A|$ of the part, which is the domain of the other component, and
(2) i_A, which is the monomorphic given inclusion of A into X.

To say that A and B are equivalent parts of X means that there is

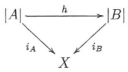

with h invertible and with the equation

$$i_B h = i_A$$

satisfied (which is denoted $A \equiv_X B$, see Definition 2.24). By contrast $|A| \cong |B|$ means merely that $|A|, |B|$ have the same number of elements but *not* necessarily in a way that respects the inclusions. We commonly say that "A is a three-element part of X" to mean that A is a part of X such that $|A|$ has three elements; two given three-element parts of X could have no elements in common, could overlap nontrivially, or even could have the same elements of X as members, but only in the last case would they be isomorphic *as parts* of X. Once this much is clearly understood, one then usually follows the "abuse of notation," which drops the $|\ |$ sign and just uses the same symbol A to stand both for the part (which involves a given i_A as understood) *or* for the domain of the part; one then sees from the context whether morphisms of parts (i.e., inclusion relation) or maps of the underlying sets are being discussed. Recall that our notion of membership is

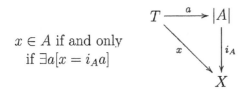

$x \in A$ if and only if $\exists a[x = i_A a]$

For any $X \xrightarrow{f} Y$ we will construct an induced map

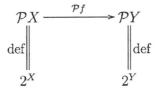

This operation arises so frequently that it has at least four other notations

$$\mathcal{P}f = f_! = f[\ ] = \exists_f = \mathrm{im}_f$$

where the $f[\ ]$ suggests that it is a generalization to parts of the evaluation of f at elements, the $\exists_f$ suggests its intimate connection with the "there exists" operation of logic, and the last im_f indicates that it is the internalization of the "geometric" operation of taking the *direct image* of a part of X. The latter operation, for given

$X \xrightarrow{f} Y$, is as follows: Given any part A of X, factorize the composite $f i_A$ into surjective and injective; the resulting part of Y is called $f[A]$. (See the picture following Exercise 8.3.)

$$
\begin{array}{ccc}
|A| & \xrightarrow{\text{surjective}} & |f[A]| \\
{\scriptstyle i_A}\Big\downarrow & & \Big\downarrow{\scriptstyle i_{f[A]}} \\
X & \xrightarrow{\quad f \quad} & Y
\end{array}
$$

The following exercise implies in particular that any two such factorizations will give *equivalent parts* of Y.

Exercise 8.1
("diagonal fill-in")

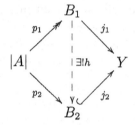

If j_2 is injective, p_1 surjective, and $j_2 p_2 = j_1 p_1$, then there exists an h (necessarily unique) for which both

$$
p_2 = h p_1
$$
$$
j_1 = j_2 h
$$

$\Diamond$

Since equivalent parts correspond to equal characteristic functions, we have a map

$$
2^X \xrightarrow{\ \text{im}_f\ } 2^Y
$$

at least if we rely on the principle that any well-defined process determines a map. Such reliance can be avoided by a method (similar to that used before) to deduce the existence of im_f as a map from a few previously assumed specific axioms, as follows:

Exercise 8.2
Let E_X denote the part of $X \times 2^X$ whose characteristic function is the *evaluation* map $X \times 2^X \xrightarrow{\epsilon_X} 2$. Show that the image I_f of E_X along the map $f \times 1_{2^X}$

$$
\begin{array}{ccc}
E_X & \longrightarrow & I_f \\
\Big\uparrow & & \Big\uparrow \\
X \times 2^X & \xrightarrow{\ f \times 1_{2^X}\ } & Y \times 2^X
\end{array}
$$

is a part of $Y \times 2^X$, whose characteristic function $Y \times 2^X \longrightarrow 2$ has the desired

$$2^X \xrightarrow{\text{im}_f} 2^Y$$

as its exponential transpose; i.e.

$$\text{im}_f \ulcorner A \urcorner = \ulcorner f[A] \urcorner$$

where the further "abuse" of notation

$$\ulcorner A \urcorner = \ulcorner \text{ characteristic function of } A \urcorner$$

has been used. ◇

The connection of direct image with existential quantification is laid bare by the following exercise:

Exercise 8.3
If in

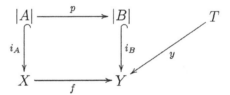

p is surjective; A, B are parts of X, Y, respectively; and $f i_A = i_B p$, then for any y

$$y \in B \iff \exists x [fx = y \text{ and } x \in A]$$

$$X \xrightarrow{\quad\quad f \quad\quad} Y$$

◇

What is called in deductive logic the "rule of inference for existential quantifier introduction and elimination" corresponds precisely to a geometric property relating inverse image and direct image. You can prove this rule in Exercise 8.5 below, but first note that

$$2^{f} \ulcorner B \urcorner = \ulcorner f^{-1} B \urcorner = Bf$$

(The notation Bf is especially apt if one thinks of B here as the *characteristic function* manifestation of the part.) If A is any part of X and B is any part of Y, there may or may not exist a map h for which

$$
\begin{array}{ccc}
|A| & \dashrightarrow^{h} & |B| \\
{\scriptstyle i_A}\downarrow & & \downarrow{\scriptstyle i_B} \qquad fi_A = i_B h \\
X & \xrightarrow{\ f\ } & Y
\end{array}
$$

(it would be unique since i_B is monic); if such a map does exist, we might write as a further definition

$$A \subseteq_f B$$

Exercise 8.4
$A \subseteq_f B \iff \forall x[x \in A \implies fx \in B]$.   ◊

Exercise 8.5
Show that

$$f[A] \subseteq_Y B$$
$$\Updownarrow$$
$$A \subseteq_X Bf$$

by demonstrating that both are equivalent to $A \subseteq_f B$, that is,

if for any x, $A(x)$ implies $B(fx)$

then for any y, $\exists x[fx = y$ and $A(x)]$ implies $B(y)$

and conversely. (Here $A(x)$ means $x \in A$; i.e., we are using the same symbol A to denote both a part and its characteristic function, which is an abuse of notation that should cause no confusion since we now understand the difference.)   ◊

Exercise 8.6
Interpret the foregoing exercise in the case $X = Y \times T$, $f = $ projection to Y.   ◊

8.2 The Covariant Power Set Functor

The covariant power set functor $\mathcal{P}$ is by definition

$$\mathcal{P}X = 2^X$$

$$\mathcal{P}f = \operatorname{im}_f$$

The functoriality as usual refers to an equation:

Exercise 8.7
For $X \xrightarrow{f} Y \xrightarrow{g} Z$ one has

$$\mathcal{P}(gf) = (\mathcal{P}g)(\mathcal{P}f)$$

as maps $\mathcal{P}X = 2^X \longrightarrow 2^Z = \mathcal{P}Z$. In other words

$$(gf)[A] = g[f[A]]$$

for any part A of X.

Hint: Use the diagonal fill-in exercise (and the fact that the composites of surjectives are surjective) to conclude that the composite on top in the following picture is the other half of the image factorization of $(gf)i_A$:

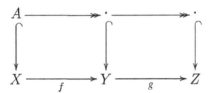

Alternatively, (introducing properties, elements, and existential quantifiers) show that for any z and for any property . . . ,

$$\exists x[(gf)x = z \text{ and } \dots] \Longleftrightarrow \exists y[gy = z \text{ and } \exists x[fx = y \text{ and } \dots]]$$

$\Diamond$

The claim that $\mathcal{P}f$,

$$A \longmapsto f[A]$$

is one generalization of the basic evaluation

$$x \longmapsto f(x)$$

on elements becomes more clear if we first internalize the idea that parts are generalized elements in a definite manner

$$X \xrightarrow{\{\}} \mathcal{P}X$$

known as the *singleton map*, which assigns to every element x the name of the characteristic function of x considered as a part of X. This map can be constructed in basic steps as follows: The diagonal map

$$X \xrightarrow{\delta_X} X \times X$$

is a part of $X \times X$ (since it is split by the projections) and hence has a characteristic map

$$X \times X \xrightarrow{\ominus_X} 2$$

where (writing $a \varphi b$ for the typical value $\varphi \langle a, b \rangle$ of a map $A \times B \xrightarrow{\varphi} C$)

$$x_1 \ominus_X x_2 = \text{true} \iff x_1 = x_2$$

But like any map whose domain is a product, $\ominus_X$ has an exponential transpose

$$X \longrightarrow 2^X$$

Exercise 8.8
If $A \times B \xrightarrow{f} C$ has the two exponential transposes $A \xrightarrow{f_1} C^B$ and $B \xrightarrow{f_2} C^A$, and if $A = B$, then

$$f_1 = f_2 \iff f(a, b) = f(b, a) \text{ for all } \langle a, b \rangle \in A \times A$$

$\diamond$

The latter condition is satisfied for $\ominus_X$. Hence, the diagonal map gives rise by the procedure above to (not two, but) one map

$$X \xrightarrow{\{\}} 2^X = \mathcal{P}X$$

such that

$$x' \in \{x\} \iff x' = x$$

Now the statement that $\mathcal{P}f$ generalizes evaluation at elements becomes the following naturality (see "natural transformation" in Appendix C.1):

Exercise 8.9
For any $X \xrightarrow{f} Y$, the square below commutes

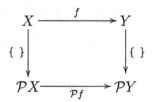

$\diamond$

In case f is *itself a monomorphism* (hence makes X the underlying set of part of Y), $\mathcal{P}f$ can be calculated without any real use of existential quantification since fi_A is already injective (hence does not need to be further factored):

$$
\begin{array}{ccc}
|A| & =\!=\!=\!=\!= & |f[A]| \\
\,\,{\scriptstyle i_A}\big\downarrow\!\!\!\!\cap & & \cap\!\!\!\!\big\downarrow \\
X & \xrightarrow{\;\;f\;\;} & Y
\end{array}
\qquad
\begin{array}{l}
|f[A]| = |A| \\
i_{f[A]} = fi_A
\end{array}
$$

Proposition 8.10: *If f is a monomorphism, then 2^f is a split epimorphism; in fact $\mathcal{P}f$ is a section for 2^f.*

Proof: For any f one has

$$f[A] \subseteq B \text{ if and only if } A \subseteq Bf = 2^f(B)$$

so in particular

$$f[A] \subseteq f[A] \text{ if and only if } A \subseteq 2^f(f[A]) = (2^f \circ \mathcal{P}f)(A)$$

i.e.,

$$A \subseteq (2^f \mathcal{P}f)(A)$$

for all parts A of X. But if f is monic, then $i_{(\mathcal{P}f)A} = fi_A$, whose inverse image $2^f(\;\;)$ is reduced to A. In other words, the square

$$
\begin{array}{ccc}
A & =\!=\!=\!=\!= & A \\
\,\,{\scriptstyle i_A}\big\downarrow\!\!\!\!\cap & & \cap\!\!\!\!\big\downarrow{\scriptstyle i_{(\mathcal{P}f)A}} \\
X & \xrightarrow{\;\;f\;\;} & Y
\end{array}
$$

is almost trivially seen *itself* to satisfy the universal mapping property of an inverse image square, that is,

$$A = 2^f((\mathcal{P}f)(A)) \text{ for all } A \text{ in } \mathcal{P}X = 2^X$$

We conclude that $2^Y \xrightarrow{\;2^f\;} 2^X$ is surjective, split by $\mathcal{P}f$. ∎

The contravariant power set functor $2^{(\;)}$ and the covariant power set functor $\mathcal{P}$ both come up in geometry and analysis in many ways, for example in connection with the condition that a map be continuous or that it be locally bounded. A continuous space (topological space) X is often considered to consist of an underlying abstract set $|X|$ of points, together with a set $\mathcal{F}_X$ of "closed parts" and a structural map

$|X| \times \mathcal{F}_X \xrightarrow{\in_X} 2$ indicating when any given point is a member of any given closed part. The transpose of $\in_X$, $\mathcal{F}_X \longrightarrow 2^{|X|} = \mathcal{P}|X|$ interprets each closed part of X as a part of the set $|X|$. Usually, special conditions are assumed on $\mathcal{F}_X$, $\in_X$, but we will not need these here (however see Exercise 8.11 for a class of examples.) On the other hand, a local boundedness space ($=$ bornological set) X involves an underlying set $|X|$ of points, a set $\mathcal{B}_X$ of "bounded parts," and a membership map $|X| \times \mathcal{B}_X \xrightarrow{\in_X} 2$; this is apparently exactly the same sort of thing $\mathcal{B}_X \longrightarrow 2^X$ as a continuous space, except for the different words. The special axiomatic conditions usually imposed on a system of bounded parts are, of course, quite different from those usually imposed on a system of closed parts (for example, an arbitrary subpart of a bounded part should again be bounded, whereas no such thing is true for closed sets; the whole part 1_X is always considered closed, but 1_X is bounded only for trivial boundedness systems). However, the striking contrast between closed and bounded is seen in the concepts of continuous versus bornological mapping. Namely, if X and Y are continuous spaces, a continuous map $X \xrightarrow{f} Y$ is a map $|X| \xrightarrow{|f|} |Y|$ of points together with a proof that the *inverse image of closed parts is closed*:

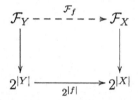

By contrast, if X and Y are local boundedness spaces, a bornological map $X \xrightarrow{f} Y$ is a map $|X| \xrightarrow{|f|} |Y|$ of points together with a proof that the *direct images of bounded parts are bounded*:

$$\mathcal{B}_X \overset{\mathcal{B}_f}{-----\!\!\!\rightarrow} \mathcal{B}_Y$$

$$\mathcal{P}|X| \xrightarrow[\mathcal{P}|f|]{} \mathcal{P}|Y|$$

Exercise 8.11

A **metric space** X involves a set $|X|$ of points and a distance function

$$|X| \times |X| \xrightarrow{d_X} \mathbb{R}_{\geq 0}^{\infty}$$

where $\mathbb{R}_{\geq 0}^{\infty} = \{r \mid 0 \leq r \leq \infty\}$ is the set of (nonnegative, extended) real numbers and d_X is required to satisfy

$$d_X(x, x) = 0$$
$$d_X(x, y) + d_X(y, z) \geq d_X(x, z)$$

for any $\langle x, y, z \rangle \in |X|^3$. This inequality can be pictured as

A map $X \overset{f}{\longrightarrow} Y$ of metric spaces is (often required to be) **distance-decreasing** in the weak sense that

$$d_Y(f x_1, f x_2) \leq d_X(x_1, x_2)$$

for any $\langle x_1, x_2 \rangle \in |X|^2$. Each metric space has a natural notion of bounded part:

A part B of $|X|$ is **bounded** if and only if it is contained in some ball with some center and some finite radius, i.e.,

$$B \in \mathcal{B}_X \overset{\text{def}}{\Longleftrightarrow} \exists x_0 \exists r[r < \infty \text{ and } \forall x[x \in B \Rightarrow d_X(x, x_0) \leq r]]$$

But each metric space also has a natural notion of **closed** part, namely, any part containing each point that can be *approximated* by points known to be in the part:

$$A \in \mathcal{F}_X \overset{\text{def}}{\Longleftrightarrow} \forall x[\forall r[r > 0 \Rightarrow \exists a[a \in A \text{ and } d_X(x, a) \leq r]] \Rightarrow x \in A]$$

Show that any map f of metric spaces (which is distance-decreasing in the weak sense) is both continuous (in the sense that 2^f carries the natural closed parts of the codomain back to natural closed parts of the domain) as well as bornological (in the sense that $\mathcal{P} f$ carries the natural bounded parts of the domain to the natural bounded parts of the codomain). $\diamond$

8.3 The Natural Map $\mathcal{P}X \longrightarrow 2^{2^X}$

Both the functor $\mathcal{P}$ and the (larger) twice-contravariant $2^{2^{(\ )}}$ are covariant; we will describe a direct natural comparison

$$\mathcal{P}X \overset{\int_X}{\longrightarrow} 2^{2^X}$$

between them (which will turn out to be injective). Given any part A of X,

$$\int_X (\ )A$$

will be a map

$$2^X \longrightarrow 2$$

namely, the one whose value at any $X \xrightarrow{\varphi} 2$ is the truth value (in $2 = \{0, 1\}$)

$$\int_{x \in X} \varphi(x) A(dx)$$

calculated as follows: if there is any x *in* A at which φ takes the value 1, then the answer is 1, whereas if φ restricted to A is constantly 0 (φ might take value 1 outside of A; that does not matter), then the answer is 0. Of course, the φ's are "really" (characteristic functions of) parts also, so the description of $\int_X$ could equally well be stated: for any given part A in $\mathcal{P}X$, consider any part φ in 2^X and ask if φ intersects A. If so the answer is 1, otherwise 0. For example,

$\int_X \varphi_1(x) A(dx) = 0$

$\int_X \varphi_2(x) A(dx) = 1$

$\int_X \varphi_3(x) A(dx) = 1$

Thus,

$$2^X \xrightarrow{\int_X (\ )A} 2$$

determines an element of 2^{2^X}, but this can be done for any part A in $\mathcal{P}X$, and hence there should be a map $\mathcal{P}X \longrightarrow 2^{2^X}$. The reason the domain is not being called 2^X will be seen in Exercise 8.13 below.

Exercise 8.12
The existence of the map $\int_X$ can be deduced from previous concrete constructions and its nature clarified for later use. Note that $2 = \mathcal{P}1$ and that there is a canonical "union" map

$$\begin{array}{ccc} \mathcal{P}\mathcal{P}1 & \xrightarrow{\mu_1} & \mathcal{P}1 \\ \| & & \| \\ \mathcal{P}2 & \longrightarrow & 2 \end{array}$$

that acts as follows:

For each of the four parts of 2,

$$\mu_1(\{0, 1\}) = \mu_1(\{1\}) = 1$$

$$\mu_1(\{0\}) = \mu_1 0 = 0$$

Moreover, the covariant functoriality of $\mathcal{P}$ (discussed in the previous sections) may be considered (further internalized) as a canonical map

$$Y^X \times \mathcal{P}X \longrightarrow \mathcal{P}Y$$
$$f, A \longmapsto f[A]$$

Hence, in particular, taking $Y = 2 = \mathcal{P}1$, we have

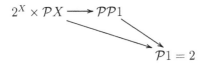

and therefore by taking the exponential transpose of the composite we have

$$\mathcal{P}X \longrightarrow 2^{2^X}$$

which is the same as our $\int_X$.   ◇

Exercise 8.13
For any $X \xrightarrow{f} Y$, the square

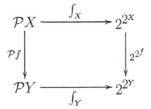

is commutative.   ◇

Exercise 8.14
Let $X \xrightarrow{\delta} 2^{2^X}$ be the evaluation map $\delta(x)(\varphi) = \varphi(x)$. Then the triangle

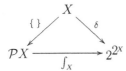

is commutative.   ◇

Now only some of the elements of 2^{2^X} come from elements of $\mathcal{P}X$ by our $\int_X$ process:

Exercise 8.15
If $2^X \xrightarrow{\alpha} 2$ is such that there is an A for which

$$\alpha(\varphi) = \int_{x \in X} \varphi(x) A(dx)$$

for all $X \xrightarrow{\varphi} 2$, then α has the following properties:

$$\alpha(0_X) = 0$$

where $0_X : X \longrightarrow 1 \xrightarrow{0} 2$ is the constantly-0 function, and

$$\alpha(\varphi_1 \cup \varphi_2) = \alpha\varphi_1 \vee \alpha\varphi_2$$

for any $X \overset{\varphi_1}{\underset{\varphi_2}{\rightrightarrows}} 2$, where

$$(\varphi_1 \cup \varphi_2)(x) \overset{\text{def}}{=} \varphi_1(x) \vee \varphi_2(x)$$

for all x, where $\vee$ is defined by $v_1 \vee v_2 = 1$ if and only if $v_1 = 1$ or $v_2 = 1$ for v_1, v_2 in 2. Give examples of maps $2^X \xrightarrow{\alpha} 2$ that do not have the two linearity properties here shown to hold for $\int_X( )A$.   ◊

Comment: Such linear α's are sometimes called "grills"; they may be considered as generalized parts in that every part A determines a grill, and the union of two grills can be defined in a way extending the union operation on parts. But there are many grills that are not parts. For example, in the plane X, we can define an α by $\alpha(\varphi) = 1$ if and only if φ is a part of the plane with positive area.

Remark: All the general constructions in Sections 8.1–8.3 (such as singleton, the map from the covariant power set to the doubly contravariant one, etc.) are applicable in any topos with 2 replaced by Ω.

8.4 Measuring, Averaging, and Winning with V-Valued Quantities

Several natural restrictions can be placed on elements of V^{V^X} to yield a covariant subfunctor

$$\text{Hom}_?(V^X, V) \lhook\joinrel\longrightarrow V^{V^X}$$

of mathematical interest. When V is equipped with some structure, as for example with $V = 2$ above, conditions similar to linearity may be imposed. But let us briefly consider two conditions that might reasonably be defined using only the knowledge that V is a given set. We want that at least the evaluation functionals $V^X \xrightarrow{\hat{x}} V$ (given by $\hat{x}(\varphi) = \varphi(x)$ for all φ (i.e., $\hat{x} = \delta(x)$)) be included, and this will be true of both our conditions.

Suppose that every month the same set X of people are to be asked which of a set V of three brands of a product they prefer, giving a result $X \xrightarrow{\varphi} V$ each month. We want to devise a single procedure α that will give $\alpha(\varphi)$ an element of V so that we can reasonably say that the people prefer $v = \alpha(\varphi)$ this month. One obvious condition that α will need to satisfy is that it should be an averaging functional to the extent that if it should happen that all people choose at φ the same brand v, that same v should surely be the result $\alpha(\varphi)$.

Definition 8.16: $V^X \xrightarrow{\alpha} V$ is **weakly averaging** *if for each v, if $\varphi_v(x) \overset{def}{=} v$ for all x, then*

$$\alpha(\varphi_v) = v$$

This is obviously an extremely weak condition and may be satisfied by functionals that nobody would want to consider as averaging. (When V has additional structure, it makes sense to require that α also be linear, in which case there are often theorems saying that a linear averaging procedure is an **expectation** with respect to a probability distribution on X.)

Proposition 8.17: *For any x in X, $V^X \xrightarrow{\hat{x}} V$ is weakly averaging. That is, if Hom_V denotes the constant-preserving functionals, one has*

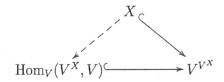

Proof: Obvious.   ■

Thus, a trivial way of defining a weakly averaging $V^X \longrightarrow V$ is to choose some single individual x in X and then evaluate every survey φ by merely noting how x answered and presenting his or her answer as the "digested result of the survey". This offends our sense of fairness to the extent that we are led to formulate a different sort of condition (which we will not study here) that could be imposed on an averaging functional:

Definition 8.18: *The arrow $V^X \xrightarrow{\alpha} V$ is said to be **symmetric** if for every invertible self-map (also called automorphism or permutation) of X, $X \xrightarrow[\sigma]{\sim} X$*

$$\alpha(\varphi\sigma) = \alpha(\varphi)$$

for all $X \xrightarrow{\varphi} V$.

We will emphasize instead a certain kind of highly skewed weakly averaging functionals. The idea is that one might hope to recover X from $\Phi = V^X$ by putting strong enough conditions on maps $\Phi \longrightarrow V$ to characterize the values $\hat{x}$ of δ in case $\Phi = V^X$. Such considerations are basic to almost all situations in algebraic geometry, particle physics, functional analysis, and so on, wherein one wants to locate or recapture points of space or states of motion X by comparing (in quantitative ways) observable variable quantities in V^X with (relatively) constant quantities in V.

Definition 8.19: *A V-generalized point of X is a functional $V^X \xrightarrow{\alpha} V$ such that for each $V \xrightarrow{\lambda} V$,*

$$\alpha(\lambda\varphi) = \lambda(\alpha(\varphi))$$

for all $X \xrightarrow{\varphi} V$, where $\lambda\varphi$ denotes the composition and $\lambda(v)$ denotes usual evaluation. That is

$$
\begin{array}{ccc}
V^V \times V^X & \xrightarrow{\ 1_{V^V} \times \alpha\ } & V^V \times V \\
{\scriptstyle \circ}\downarrow & & \downarrow{\scriptstyle \text{eval}} \\
V^X & \xrightarrow{\quad \alpha \quad} & V
\end{array}
$$

should commute.

This condition is quite strong. For example, if V denotes the real numbers and α is an expectation, and if we consider as an example of λ the operation of multiplying by a constant (e.g., doubling), then

$$\alpha(\lambda\varphi) = \lambda(\alpha(\varphi))$$

follows from linearity. But instead take λ to be squaring: just the case

$$\alpha(\varphi^2) = \alpha(\varphi)^2$$

of our condition forces (the distribution underlying) α to be so concentrated that the standard deviation (relative to α) of all random variables φ is zero!

Proposition 8.20: *Each point is a generalized point.*

Proof: If $\alpha = \hat{x}$, then for any $X \xrightarrow{\varphi} V \xrightarrow{\lambda} V$

$$\alpha(\lambda\varphi) = (\lambda\varphi)x = \lambda(\varphi x) = \lambda(\alpha(\varphi))$$

∎

Exercise 8.21
Any generalized point is weakly averaging.

Hint: Each element of V determines a constant self-map of V, $V \lhook\joinrel\longrightarrow V^V$, and this can be used to show that

$$X \lhook\joinrel\longrightarrow \text{Hom}_{V^V}(V^X, V) \lhook\joinrel\longrightarrow \text{Hom}_V(V^X, V) \lhook\joinrel\longrightarrow V^{V^X}$$

where Hom_{V^V} denotes the generalized points, that is, the functionals α that are homogeneous with respect to all λ in V^V. ◊

To what extent all generalized points are really just the original elements has long been of interest. Isbell showed in 1960 [Is60] that if V is a given infinite set, then

$$X \xrightarrow{\sim} \text{Hom}_{V^V}(V^X, V)$$

for all sets X that arise in ordinary geometry and analysis even though by Cantor's theorem most of these X's are larger than V. However, many set-theorists study the possibility of the existence of sets X so large that they cannot be "measured" by any fixed V in this way (paradoxically, these hypothetical extremely large sets are usually called "measurable cardinals").

For a small V, like $V = 2$, it is easy to find examples of generalized points that are not points but are interesting in view of some applications: V^V has four elements λ, of which the identity 1_V contributes no restriction on generalized points. Thus,

Proposition 8.22: *A functional* $2^X \xrightarrow{\alpha} 2$ *commutes with all* λ *in* 2^2 *if and only if*

$$\alpha(\ulcorner X \urcorner) = 1, \quad \alpha(0) = 0$$

$$\alpha(\varphi) = 1 \iff \alpha(\neg\varphi) = 0$$

(where $\neg : 2 \longrightarrow 2$ *is defined by* $\neg 0 = 1, \neg 1 = 0$*).* ∎

Such generalized points α are sometimes considered to arise from a context in which the elements of X can "choose sides" in various ways to produce an alignment φ that will either "win" $\alpha(\varphi) = 1$ or "lose" $\alpha(\varphi) = 0$. (The following is a further restriction on α sometimes considered: if $\varphi \subseteq \varphi'$ and $\alpha(\varphi) = 1$ then $\alpha(\varphi') = 1$. However, it is not true in all situations α that a team φ' bigger than a winning team φ would also win (consider what may happen if too many CIA and MI5 individuals infiltrate our team)). We want to see that there are situations α that are generalized points of X in the sense that they preserve the action of all $\lambda \in 2^2$ but are not points $\hat{x}_0$; in other words, the situation α is more complex than one in which there is such a towering star player x_0 that for any φ, φ is a winning team if and only if x_0 is on φ. If X is too small, such complexity will not be possible.

Exercise 8.23

If $X = V$, then $X \xrightarrow{\sim} \text{Hom}_{V^V}(V^X, V)$. Take the example $\varphi = 1_X$ but note that in this case a condition λ in V^V may be considered also as an arbitrary input for a functional. ◊

Thus, for $V = 2$, we need to take X with at least three elements to find generalized points α which do not come from points.

Exercise 8.24

If X has n elements, show that the number of maps $2^X \xrightarrow{\ \alpha\ } 2$ that commute with the action of all four λ in 2^2 is

$$2^{2^{n-1}} - 1$$

For the case $n = 3$, $X = \{a, b, c\}$, determine explicitly all eight α's by displaying, for each α all the teams φ that win α.

(Such φ can conveniently be displayed as subsets of $\{a, b, c\}$.) $\qquad \Diamond$

Exercise 8.25

Let $V = 3$. Show that the 27 conditions a generalized point must satisfy all follow from a much smaller number among them. Show that any generalized point of any finite set X is in fact a point of X itself. $\qquad \Diamond$

Remark 8.26: The recovery of X from measurement, i.e.,

$$X \xrightarrow{\ \sim\ } \mathrm{Hom}_{V^V}(V^X, V)$$

is also valid when X is "any" metrizable continuous space, $V = \mathbb{R}$ (the space of real numbers), and all maps are continuous. $\qquad \Diamond$

8.5 Additional Exercises

Exercise 8.27

To define the covariant power set functor we used only the elementary (topos) properties of $\mathcal{S}$. Describe im_f for an arrow $X \xrightarrow{\ f\ } Y$ in $\mathcal{S}/X$. Note that to do this it is first necessary to describe the objects $\mathcal{P}X$ (recall Exercises 5.12, 6.8).

Exercise 8.28

Describe im_f for an arrow $X \xrightarrow{\ f\ } Y$ in $\mathcal{S}^{2^{\mathrm{op}}}$.
Recall Section 6.2 and Exercise 6.13.

Exercise 8.29

Describe im_f for an arrow $X \xrightarrow{\ f\ } Y$ in M-sets.
Recall Exercises 5.13, 6.15.

Exercise 8.30

Similarly, to define the singleton arrow, $X \xrightarrow{\{\}} \mathcal{P}X$, we again used only the elementary (topos) properties of $\mathcal{S}$. Describe $X \xrightarrow{\{\}} \mathcal{P}X$ for an object X in each of $\mathcal{S}/A$, $\mathcal{S}^{2^{\text{op}}}$ and M-sets. Keep in mind that 2 in the construction of Section 8.2 must be replaced by the appropriate truth-value object.

Exercise 8.31

Show that the diagonal arrow $X \xrightarrow{\Delta_X} X \times X$ provides components of a *natural transformation* (see Appendix C.1) from the identity functor to the composite functor $(- \times -)\Delta$ (see Exercise 5.17). Find a natural transformation from the composite functor $\Delta(- \times -)$ to the identity functor on $\mathcal{S} \times \mathcal{S}$.

9

Introduction to Variable Sets

9.1 The Axiom of Infinity: Number Theory

Recall that we denote by $\mathcal{S}$ the category of (abstract, discrete, constant) sets and *arbitrary* maps between them that we have studied till now. In the various branches of mathematics (such as mechanics, geometry, analysis, number theory, logic) there arise many different categories $\mathcal{X}$ of (not necessarily discrete, variable) sets and *respectful* maps between them. The relation between $\mathcal{S}$ and the $\mathcal{X}$'s is (at least) threefold:

(0) $\mathcal{S}$ is "case zero" of an $\mathcal{X}$ in that in general the sets in $\mathcal{X}$ have some sort of structure such as glue, motion and so on, but in $\mathcal{S}$ this structure is reduced to nothing. However, the general $\mathcal{X}$ often has a functor $\mathcal{X} \xrightarrow{\;| \;|\;} \mathcal{S}$ determining the mere number (Cantor) $|X|$ of each such emergent aggregate X.

(1) A great many of the mathematical properties of such a category $\mathcal{X}$ of variable sets are the same or similar to properties of the category $\mathcal{S}$ of constant sets. Thus, a thorough knowledge of the properties of $\mathcal{S}$, together with some categorical wisdom, can be indispensable in dealing with problems of analysis, combinatorics, and so forth. The main common properties include the concepts of function spaces X^T and of power sets $\mathcal{P}(X)$.

(2) Many examples of categories of variable sets $\mathcal{X}$ are reconstructible from $\mathcal{S}$ as $\mathcal{X} = \mathcal{S}^{\mathbf{T}^{\mathrm{op}}}$, where $\mathbf{T}$ is a datum (which can also be described in $\mathcal{S}$) whose role is to specify the general nature of the glue, motion, and so on, in which all sets in $\mathcal{X}$ participate, and $\mathcal{X}$ is considered to consist of all possible examples having that particular general nature. (Sometimes such a datum $\mathbf{T}$ is called a **theory**; we will explicitly consider a few such theories, some of which can actually be extracted in a fairly simple way from $\mathcal{X}$ itself.)

The connection between $\mathcal{X}$ and $\mathcal{S}^{\mathbf{T}^{\mathrm{op}}}$ often comes about in the following way: among the variable sets in $\mathcal{X}$ there are a few special ones $\mathbf{T} \xhookrightarrow{\quad} \mathcal{X}$ such that knowledge of the $|X^T|$ for T in $\mathbf{T}$ *and* their interactions as *constant* sets (in $\mathcal{S}$)

154

suffices to reconstruct any variable set X. ($\mathbf{T} = \uparrow$ was considered in Chapter 6.) A basic example of a mode of variability of sets is $\mathbf{T} = \boxed{\circlearrowleft}$. That is, we consider a category $\mathcal{S}^{\circlearrowleft}$ of variable sets X in which X has not only elements but also an internal "dynamic" of the sort that any given element uniquely determines a "next" element, and the maps between sets respect this internal dynamic. The category $\mathcal{S}^{\circlearrowleft}$ can be regarded as constructed from $\mathcal{S}$ as follows:

A set X of $\mathcal{S}^{\circlearrowleft}$ "is" (i.e., determines and is determined by) a set of $\mathcal{S}$ together with an endomap as structure

$$X^{\circlearrowleft \xi_X} \text{ in } \mathcal{S}$$

A map of $\mathcal{S}^{\circlearrowleft}$ "is" two sets $X^{\circlearrowleft \xi_X}, Y^{\circlearrowleft \xi_Y}$ and a mapping f in $\mathcal{S}$ satisfying the equation in $\mathcal{S}$,

$$f \xi_X = \xi_Y f$$

so the following diagram commutes:

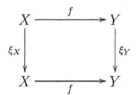

Composition in $\mathcal{S}^{\circlearrowleft}$ of two maps $X^{\circlearrowleft \xi_X} \xrightarrow{f} Y^{\circlearrowleft \xi_Y} \xrightarrow{g} Z^{\circlearrowleft \xi_Z}$ is only defined in case the codomain of f and the domain of g (not only have the same underlying abstract set of elements but also) have the same internal dynamic ξ_Y. In that case the composite is formed by forming the gf as in $\mathcal{S}$ and forgetting ξ_Y while retaining ξ_X and ξ_Z. Note that we will sometimes abuse notation and not mention the dynamic ξ_X of $X^{\circlearrowleft \xi_X}$.

Exercise 9.1

If $f \xi_X = \xi_Y f$ and if $g \xi_Y = \xi_Z g$, then $(gf) \xi_X = \xi_Z (gf)$. That is, the composite defined above really does give a result that is again a map in $\mathcal{S}^{\circlearrowleft}$. ◊

Exercise 9.2

Let 1 denote a one-element set equipped with the only possible dynamic (endomap). For any X in $\mathcal{S}^{\circlearrowleft}$, there is a unique $\mathcal{S}^{\circlearrowleft}$-map $X \longrightarrow 1$. Show also that $\mathcal{S}^{\circlearrowleft}$ has an initial object. ◊

Exercise 9.3

A map $1 \xrightarrow{x} X$ in $\mathcal{S}^{\circlearrowleft}$ is always the same thing as a *fixed point* of the dynamic of X, that is, an element x of the underlying abstract set of X for which

$$\xi_X x = x$$

◊

Thus, the maps from $T = 1$, although an extremely important aspect of the sets X in $S^{\circlearrowleft}$, fall far short of detecting all the elements of the underlying set of X; another object $T = N$ is necessary for that purpose, and that will force the existence in S itself of *infinite* sets. In more detail, the underlying abstract set of N (also denoted N) must be equipped with an S-endomap $N^{\circlearrowleft\sigma}$ (usually called "successor") of the underlying set (in order to be a set of $S^{\circlearrowleft}$ at all), and we want there to be a natural invertible correspondence

$$\frac{N \xrightarrow{\ x\ } X \text{ in } S^{\circlearrowleft}}{1 \xrightarrow{\ x_0\ } X \text{ in } S} \quad \uparrow\downarrow$$

Taking the case $X = N$, $x = 1_N$, we deduce that there must be a distinguished element

$$1 \xrightarrow{\ x_0 = 0\ } N$$

of the underlying abstract set of N. Since just to say $N \xrightarrow{\ x\ } X$ in $S^{\circlearrowleft}$ is to say that $x\sigma = \xi_X x$ in S, we are led to the following statement characterizing the system

$$1 \xrightarrow{\ 0\ } N^{\circlearrowleft\sigma}$$

of maps in $S^{\circlearrowleft}$ in terms of S only:

AXIOM: DEDEKIND–PEANO
There exist $1 \xrightarrow{\ 0\ } N \xrightarrow{\ \sigma\ } N$ *in S such that for any diagram*

$$1 \xrightarrow{\ x_0\ } X^{\circlearrowleft\xi}$$

in S there exists a unique sequence x for which both $x0 = x_0$ and $x\sigma = \xi x$

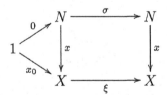

Here "sequence" is the standard name for maps in S with domain N. The equation expressed by the commutativity of the square is the condition that x be a respectful map in variable sets, and the triangular equation expresses that the element x_0 is being represented by x. Notice also that the axiom is a statement *about* S. It is precisely the *axiom of infinity* we promised in Chapter 6. With this axiom we have completed our listing of axioms for S. In any category, an object N (or more precisely, the diagram $1 \xrightarrow{\ 0\ } N \xrightarrow{\ \sigma\ } N$) satisfying the Dedekind–Peano axiom is called a **natural number object**.

Now the claim that N is not finite can be justified by examples of X. For example, one variable set in $\mathcal{S}^{\circlearrowleft}$ is

$$X_5 : \boxed{0 \quad 1 \quad 2 \quad 3 \quad 4}$$

i.e. $\xi(x) = x + 1 \quad x < 4; \qquad \xi(4) = 4$

Hence, there must exist a map $N \xrightarrow{\; x \;} X_5$ in $\mathcal{S}^{\circlearrowleft}$ with this five-element codomain, that is, a map in $\mathcal{S}$ for which

$$x0 = 0$$
$$x\sigma 0 = x0 + 1 = 1$$
$$x\sigma\sigma 0 = x\sigma 0 + 1 = 1 + 1 = 2$$
$$x\sigma\sigma\sigma 0 = x\sigma\sigma 0 + 1 = 2 + 1 = 3$$
$$x\sigma\sigma\sigma\sigma 0 = x\sigma\sigma\sigma 0 + 1 = 3 + 1 = 4$$
$$x\sigma\sigma\sigma\sigma\sigma 0 = \xi(4) = 4$$

Hence, $0, \sigma 0, \sigma\sigma 0, \sigma\sigma\sigma 0, \sigma\sigma\sigma\sigma 0$ must be distinct elements of N since they have distinct values under at least one map x. Of course under this x all the elements $\sigma^5 0, \sigma^6 0, \ldots$ get the same value in X_5. But the single set N is supposed to work for all X, so one could take $X = X_6$ to see that there is at least one x that distinguishes $\sigma^5 0$ from $\sigma^6 0$, and so on. Thus, in N all elements obtained from 0 by successive application of the successor map σ are distinct. If N had other elements than those so obtained, an X could be constructed for which two mappings $N \rightrightarrows X$ represent the same element of X, contradicting the uniqueness part of the universal mapping property of N. Thus, we are led to the following intuitive idea:

$$N = \{0, 1, 2, 3, \ldots\} \quad \sigma n = n + 1$$

9.2 Recursion

The universal property of the successor σ says (using the usual notation $x_n = xn$ for the values of a sequence) that for any given "next step" rule $X \overset{\xi}{\circlearrowleft}$ on any set X and any given starting value $1 \xrightarrow{\; x_0 \;} X$, there is a unique sequence x in X such that

$$x_0 = \text{the given starting value}$$
$$x_{n+1} = \xi(x_n) \text{ for all } n$$

The sequence x is said to be defined by simple recursion from x_0, ξ. We will prove in Lemma 9.5 the existence of sequences defined by somewhat more general kinds of recursions as well. First note that we can formally prove one of the characteristic properties of infinite sets.

Theorem 9.4: $1 + N \xrightarrow{\sim} N$. *The successor map is injective but not surjective, therefore N is Dedekind infinite.*

Proof: By the universal mapping property of coproduct there is a unique f for which

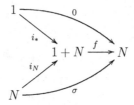

We want to define an inverse $N \xrightarrow{g} 1 + N = X$ for f by recursion. But the "recursion" satisfied by g is

$$g0 = i_*$$
$$g(n + 1) = i_N(n)$$

in which the right-hand side is not simply some function ξ of the "previous value" $g(n)$ but instead is some function of n and does not mention $g(n)$. Even dependence on both n and $g(n)$ still permits recursive definition:

Lemma 9.5: *(Recursion Lemma) If* $N \times A \xrightarrow{h} A$ *is any given map and if* $1 \xrightarrow{a_0} A$ *is any given element, then there exists a unique sequence* $N \xrightarrow{g} A$ *for which*

$$g(0) = a_0$$
$$g(n + 1) = h(n, g(n))$$

Proof: (The discrete version of a standard method for dealing with nonautonomous differential equations: augment the state space to include also the time coordinate). Define

$$X = N \times A, \quad \xi_X(n, a) = \langle n + 1, h(n, a) \rangle, \quad x_0 = \langle 0, a_0 \rangle$$

Then by simple recursion there is a unique x for which

$$x(0) = x_0 = \langle 0, a_0 \rangle$$
$$x(n + 1) = \xi x(n)$$

But if we define u and g by

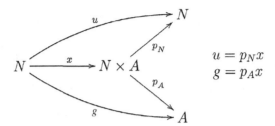

$$u = p_N x$$
$$g = p_A x$$

so that $xn = \langle un, gn \rangle$ for all n, then we have

$$\langle u0, g0 \rangle = \langle 0, a_0 \rangle$$
$$\langle u(n+1), g(n+1) \rangle = \xi \langle u(n), g(n) \rangle$$
$$= \langle u(n)+1, h(n, gn) \rangle$$

by definition of ξ. Hence, by taking projections

$$u(0) = 0$$
$$u(n+1) = u(n)+1$$

that is, $u = 1_N$ by uniqueness, and

$$g(0) = a_0$$
$$g(n+1) = h(n, gn)$$

as required; g is uniquely determined since x is. $\blacksquare$

Corollary 9.6: *There is a unique map $N \xrightarrow{p} N$ called the predecessor map for which*

$$p(0) = 0$$
$$p\sigma = 1_N$$

In other words $p(n+1) = n$ for all n. (Hence σ is injective since it has a retraction.)

Proof: Take the first projection from $N \times N$ for h in the lemma. This is the opposite extreme (from the original axiom) of the lemma, namely the rule $N \times A \longrightarrow A$ (with $N = A$) for computing the next value depends only on N rather than on A. $\blacksquare$

Now we can complete the proof of the theorem $1 + N \xrightarrow[f]{\sim} N$ because the inverse g can be defined by

$$g(0) = i_*$$
$$g(n+1) = i_N(n)$$

(similar to, but not identical with, the predecessor function). Then

$$fg(0) = 0$$
$$(fg)(n+1) = fi_N(n)$$
$$= \sigma(n) = n+1$$

so that $fg = 1_N$ by uniqueness of functions defined by recursion. Also

$$gfi_* = g(0) = i_*$$
$$gfi_N = g\sigma = i_N$$

so that $gf = 1_{1+N}$ by the uniqueness of maps from coproduct sets. Hence, $1 + N \xrightarrow{\sim} N$. ∎

9.3 Arithmetic of N

We also want to show that $N \times N \cong N$, but we need first to develop some of the arithmetic of N to be able to define the comparison maps.

Much of the arithmetic of N is implicit in the internalization of the recursion process itself. First note that

Proposition 9.7: *Given any map $A \xrightarrow{\alpha} A$ there is a sequence $N \xrightarrow{\alpha^{()}} A^A$ such that*

$$\alpha^0 = 1_A$$
$$\alpha^{n+1} = \alpha\alpha^n \quad (composition)$$

(note the abuse of notation which ignores ⌜ ⌝).

Proof: Apply the Dedekind–Peano axiom to the case where $X = A^A$, $x_0 = \ulcorner 1_A \urcorner$, and where the "next stage" endomap

$$A^A \xrightarrow{\xi} A^A$$

is the one that composes *any* map β with α: $\xi(\beta) = \alpha\beta$ for all $1 \xrightarrow{\beta} A^A$. ∎

Internalizing the proposition, we get

Theorem 9.8: *For any set A there is a map*

$$A^A \xrightarrow{iter_A} (A^A)^N$$

that assigns to any α the (name of the) sequence of iterates of α.

Exercise 9.9
Show directly that iter$_A$ exists. ◊

For example, if $A = N$ we have a procedure

$$N^N \xrightarrow{\text{iter}_N} (N^N)^N \cong N^{N \times N}$$

which assigns, to each map $N \xrightarrow{\alpha} N$, a *binary operation* $\overline{\alpha}$ on N satisfying

$$\overline{\alpha}(0, m) = m \text{ for all } m$$
$$\overline{\alpha}(n + 1, m) = \alpha(\overline{\alpha}(n, m)) \text{ for all } n, m$$

For example, if we take $\alpha = \sigma$, the basic successor operation itself, then the iteratively derived binary operation satisfies

$$\overline{\sigma}(0, m) = m$$
$$\overline{\sigma}(n + 1, m) = \sigma(\overline{\sigma}(n, m)) = \overline{\sigma}(n, m) + 1$$

This proves the existence (as a map) of the operation of *addition* of natural numbers

$$\overline{\sigma}(n, m) = n + m$$

in which notation the above recursion rules for generating $\overline{\sigma}$ become the familiar

$$0 + m = m$$
$$(n + 1) + m = (n + m) + 1$$

Diagonalizing the addition map

$$N \xrightarrow{\delta} N \times N \xrightarrow{+} N$$

we get the map $N \longrightarrow N$, which is usually denoted by $2 \cdot (\ )$, "multiplication by 2". The iteration process can be applied to the latter to yield the binary operation $\overline{\alpha}$ for which

$$\overline{\alpha}(0, m) = m$$
$$\overline{\alpha}(n + 1, m) = 2 \cdot \overline{\alpha}(n, m)$$

Exercise 9.10
Determine the operation $\overline{\alpha}$ satisfying the preceding recursion relations. ◊

Exercise 9.11
Use the maps $2 \cdot (\ )$ and $2 \cdot (\ ) + 1$ to set up an explicit invertible map
$N + N \xrightarrow{\sim} N$. ◊

To obtain the existence of multiplication as a map, note that it should satisfy the recursion

$$0 \cdot m = 0$$
$$(n + 1) \cdot m = n \cdot m + m$$

More generally, given any set A equipped with an "addition" (assumed associative) $A \times A \xrightarrow{+} A$ and a "zero" $1 \xrightarrow{0} A$, we can simply apply the iterator to the transpose of $+$ and follow the result by evaluation at 0

$$A \xrightarrow{\ulcorner + \urcorner} A^A \xrightarrow{\text{iter}} (A^A)^N \xrightarrow{ev_0^N} A^N$$

obtaining a map whose transpose is the map

$$N \times A \dashrightarrow A$$

which is usually called *multiplication*.

The idea for proving $N \times N \xrightarrow{\sim} N$ is based on the observation that the fibers of the map $N \times N \xrightarrow{+} N$ are finite (in fact the nth fiber has length $n + 1$), and so it should be possible to define a map $N \longrightarrow N \times N$ that runs through each fiber one after another, thus eventually running through all of $N \times N$ (without repetitions):

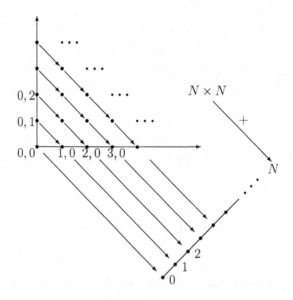

The inverse $N \times N \xrightarrow{f} N$ of the desired enumeration $N \longrightarrow N \times N$ will indicate the number of steps that have been traversed in the enumerated path in reaching a given point $\langle x, y \rangle$. The fiber of $+$ in which $\langle x, y \rangle$ lies is determined by $x + y = n$, and at the top (beginning) of the nth fiber the path has taken

$$\sum_{k<n}(k+1) = \frac{n(n+1)}{2} = \frac{(x+y)(x+y+1)}{2}$$

steps. Thus, at the point $\langle x, y \rangle$ itself, where x additional (downward to the right) steps have been taken, we get

$$f(x, y) = \frac{(x+y)(x+y+1)}{2} + x$$

This quadratic expression is actually injective and surjective when considered with only natural numbers in its domain and codomain, $N \times N \xrightarrow{f} N$. The best proof of that is to recursively define the inverse map $N \xrightarrow{\langle x,y \rangle} N \times N$ whose components are often called "pairing functions".

Exercise 9.12
Show that for any set A equipped with a commutative and associative operation $A \times A \xrightarrow{+} A$, with zero $1 \xrightarrow{0} A$, there is a map

$$A^N \xrightarrow{\sum} A^N$$

that assigns to any sequence a its sequence of partial sums

$$\left(\sum a\right)_n = \sum_{k=0}^{n-1} a_k$$

(note that $\left(\sum a\right)_0 = 0$ independent of a). ◊

The verification that, for example, the composition of two sequences defined by recursion is equal to another given sequence defined by recursion can often be done just by verifying that the effective recursion data are the same. But more generally, such equality statements are usually proved by the method known as induction. Note that the equalizer of two sequences is a part of N, therefore the usual statement of induction involves parts of N rather than equations.

Theorem 9.13: *(also known as Peano's Postulate). If A is any part of N, one can conclude that A is actually the whole of N, $A \xrightarrow[i_A]{\sim} N$, provided one can show that both*

$$0 \in A \quad and$$
$$\forall n \, [n \in A \Rightarrow n+1 \in A]$$

Proof: We need to recursively define an inverse $x : N \longrightarrow A$ for the inclusion i_A of A in N. By the second part of the hypothesis there is a map $A \xrightarrow{\sigma_A} A$ for which

$i_A \sigma_A = \sigma i_A$. Since, moreover, $0 \in A$, there is by recursion a unique $N \xrightarrow{x} A$ for which

$$x0 = 0$$
$$x\sigma = \sigma_A x$$

Then,

$$i_A x0 = 0$$
$$i_A x\sigma = i_A \sigma_A x$$
$$= \sigma i_A x$$
$$= \sigma(i_A x)$$

Therefore,

$$i_A x = 1_N$$

as the unique solution of the recursion problem whose solution is 1_N. Furthermore, we have $i_A(x i_A) = (i_A x)i_A = 1_N i_A = i_A 1_A$, and so $x i_A = 1_A$ since i_A is monic. Combining these shows that $i_A : A \xrightarrow{\sim} N$, as claimed. ∎

Now the equation

$$i_A \sigma_A = \sigma i_A$$
$$i_A \text{ monic}$$

means that $A \circlearrowleft^{\sigma_A}$ is a *part* of $N \circlearrowleft^{\sigma}$ *in the sense of the category* $\mathcal{S}^{\circlearrowleft}$ (see Exercise 9.18). The induction theorem says that if such a part contains the element 0 in its underlying abstract set, then it can only be the whole N. But how many parts of N (in the sense of $\mathcal{S}^{\circlearrowleft}$) are there altogether? There is of course the empty part. But, moreover, given any element of N, for example 5, the representing map

$$N \xrightarrow{5+()} N$$

is a part of N (also in $\mathcal{S}^{\circlearrowleft}$) whose members are $5, 6, 7, 8 \ldots$. It should be clear that there are no other parts of N in $\mathcal{S}^{\circlearrowleft}$ (even though in $\mathcal{S}$ there is a huge number 2^N of parts of N). This knowledge of the structure of parts of N will be essential in understanding the nature of a different remarkable variable set $\mathcal{P}_N(1)$, the set of truth values for $\mathcal{S}^{\circlearrowleft}$.

Exercise 9.14
Determine the classifier $\mathcal{P}_N(1)$ for sub dynamical systems.

Hint: Let 0 stand for "true" ("already true") and ∞ for "false" (i.e., "will never become true") and remember that the indicator map $X \longrightarrow \mathcal{P}_N(1)$ of any part A of any X must be a map in $\mathcal{S}^{\circlearrowleft}$, whereas the internal dynamic of $\mathcal{P}_N(1)$ must be given once and for all independently of X. Show that every map $N \longrightarrow \mathcal{P}_N(1)$ has finite image. $\diamond$

9.4 Additional Exercises

Exercise 9.15

Why does the statement $1 + N \cong N$ imply that successor is not surjective?

Exercise 9.16

Show that the addition mapping $N \times N \xrightarrow{+} N$ satisfies

$$(n + m) + p = n + (m + p)$$
$$n \mid m = m + n$$
$$n + p = m + p \Rightarrow n = m$$

Exercise 9.17

Show that the multiplication mapping $N \times N \longrightarrow N$ satisfies

$$(n \cdot m) \cdot p = n \cdot (m \cdot p)$$
$$n \cdot m = m \cdot n$$
$$n \cdot p = m \cdot p \Rightarrow n = m \text{ or } p = 0$$

Exercise 9.18

Identify the monomorphisms in $\mathcal{S}^{\circlearrowleft}$.

Exercise 9.19

Show that $\mathcal{S}^{\circlearrowleft}$ has products and equalizers (and hence all finite limits). Show that $\mathcal{S}^{\circlearrowleft}$ has all finite colimits.

Exercise 9.20

Show that $\mathcal{S}^{\circlearrowleft}$ has exponential objects. Notice that this exercise combined with the preceding exercise and Exercise 9.14 implies that $\mathcal{S}^{\circlearrowleft}$ is a topos.

Exercise 9.21

Show that there is an object in $\mathcal{S}/X$ that has an element 0 and an endomorphism σ (both in the sense of $\mathcal{S}/X$) that satisfies the Dedekind–Peano axiom.

Hint: The object is constant as a family, that is, as a mapping to X it is a projection.

Exercise 9.22

Show that $S^{2^{op}}$ has a natural number object.

Hint: The object has an identity as its transition arrow.

Exercise 9.23

Show that the category of M-sets has a natural number object.

Hint: The object has trivial action by M.

Exercise 9.24

Show that $S^{\circlearrowright}$ (!) has a natural number object.

Hint: The object has an identity as its structural endomorphism.

Exercise 9.25

(a) Show that there is a (*diagonal*) functor (Appendix C.1) denoted Δ from S to $S^{\circlearrowright}$ whose value at a set A is the object of $S^{\circlearrowright}$ given by A with the identity endomorphism. (Define Δ also on mappings and show it satisfies the equations.)

(b) Show that there is a functor Fix from $S^{\circlearrowright}$ to S that sends an object X of $S^{\circlearrowright}$ to its set of fixed points (see Exercise 9.3).

   These two functors have the following important relationship:

(c) Show that for any set A in S and object X in $S^{\circlearrowright}$ there is a one–one correspondence between mappings in $S^{\circlearrowright}$ from $\Delta(A)$ to X and mappings in S from A to Fix(X). (Δ is *left adjoint* – see Appendix C.1 – to Fix.)

10

Models of Additional Variation

10.1 Monoids, Posets, and Groupoids

The section title names three very special classes of categories that arise frequently. Knowledge of these categories is helpful in analyzing less special situations.

Definition 10.1: *A* **monoid** *is a category with exactly one object.*

Hence any two maps in a monoid $\mathcal{A}$ can be composed; this composition is associative and has a two-sided neutral element usually called 1. A monoid $\mathcal{A}$ is often said to **arise concretely** if we are given some large category $\mathcal{E}$ (of structures or spaces), if we choose an object A of $\mathcal{E}$, and if we define $\mathcal{A}$ to consist of *all* $\mathcal{E}$-endomaps of A.

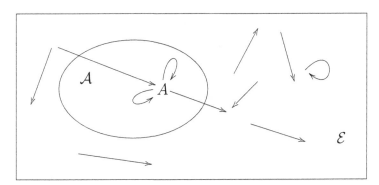

Thus, for example, all *linear* endomaps of the particular linear space $A = \mathbb{R}^3$, i.e. all 3×3 real matrices, constitute a monoid $\mathcal{A}$, where composition is matrix multiplication. Here $\mathcal{E}$ could be the concrete category of all $\mathbb{R}$-linear spaces and maps; hence, $\mathcal{A}$ can be considered to arise concretely in the preceding sense. We will study in more detail in Section 10.3 another example, the concrete monoid of all endomaps (in the sense of the concrete category $\mathcal{E} = \mathcal{S}$ of abstract sets and arbitrary maps) of the particular two-element set $A = 2$; the resulting monoid $\mathcal{A}$ has exactly four maps in it, one of which is a non-identity **involution** ($\tau\tau = 1$) and two of which are *constant* ($cx = c$ for all x in $\mathcal{A}$); moreover,

$$\tau c_0 = c_1, \tau c_1 = c_0$$

where c_0, c_1 are the two constants.

Definition 10.2: *In a category (not necessarily a monoid), the notion of* **constant map** $A \xrightarrow{c} B$ *can be defined as follows: for every object T there is a map $T \xrightarrow{c_T} B$ such that for all $T \xrightarrow{x} A$, $cx = c_T$.*

Note that in any category with a terminal object, a constant in the sense of Definition 1.25 satisfies this definition also.

Exercise 10.3
Show that $2 \xrightarrow{c_1} 2$ is constant in the sense of this definition even when we consider it within the whole category $\mathcal{S}$.   ◊

On the other hand, monoids often arise non-concretely:

Exercise 10.4
With each real number s associate "abstractly" a map T_s and define

$$1 = T_0$$
$$T_s T_t = T_{s+t}$$

Show that a monoid (i.e., a one-object category) is thus obtained. Can this monoid be realized concretely, as the full subcategory of *all* endomaps of a suitable object A in the category of all structured spaces of some kind?   ◊

Exercise 10.5
For pairs $\langle a, b \rangle$ of real numbers, define

$$\langle a_2, b_2 \rangle \cdot \langle a_1, b_1 \rangle = \langle a_2 a_1, a_2 b_1 + b_2 \rangle$$

Prove that this is a one-object category. Which pair $\langle a_0, b_0 \rangle$ is the identity map 1 of this category? Can it be concretely realized?   ◊

A second special class of categories is described by

Definition 10.6: *A category* $\mathcal{A}$ *is a* **poset** *(short for* **partially ordered set***) if and only if there is at most one map from any object to any other object:*

$$A' \underset{g}{\overset{f}{\rightrightarrows}} A \implies f = g$$

We often write $A' \leq A$ (or $A' \subseteq A$ or $A' \vdash A$) to mean

$$\exists f [A' \longrightarrow A]$$

Sometimes it is important (though usually not) to distinguish between posets in general and "strict" posets in particular, the latter being those for which the further condition

$$A' \leq A \text{ and } A \leq A' \implies A' = A$$

holds.

A poset $\mathcal{A}$ is often said to **arise concretely** if we are given a large category $\mathcal{E}$ of structured spaces, choose an object X in $\mathcal{E}$, and define $\mathcal{A}$ to be a category of parts of X (in the sense of $\mathcal{E}$) with inclusions as maps.

Exercise 10.7

In a poset all maps are monomaps. If $\mathcal{A}$ is a category that has a terminal object and in which all maps are monomaps, then $\mathcal{A}$ is a poset. Give an example of a category in which all maps are monomaps but which is not a poset. ◇

Exercise 10.8

In any category $\mathcal{B}$, we could define $B' \leq B$ to mean $\exists B' \longrightarrow B$. Explain how this idea associates to every category $\mathcal{B}$ a poset $\mathcal{B}_{po}$ and define a functor (see Definition 10.18) $\mathcal{B} \longrightarrow \mathcal{B}_{po}$. When is this functor invertible? ◇

Exercise 10.9

Consider triples $\langle x, y, r \rangle$, where x, y are arbitrary real numbers but r is real with $r > 0$. Define $\langle x, y, r \rangle \overset{f}{\longrightarrow} \langle \overline{x}, \overline{y}, \overline{r} \rangle$ to mean that $r \leq \overline{r}$ as real numbers and f is a nonnegative real number $f \geq 0$ for which

$$(\overline{x} - x)^2 + (\overline{y} - y)^2 = (\overline{r} - r + f)^2$$

Define composition so as to make this a category. Is it a poset? Can it be concretely realized? ◇

Exercise 10.10
There is only one category that is both a monoid and a poset, and it is a "groupoid" in the following sense: ◊

Definition 10.11: *A category is said to be a* **groupoid** *if and only if every arrow in it has a two-sided inverse in the same category:*

$$\forall A', A \forall A' \xrightarrow{f} A \exists A \xrightarrow{g} A'[fg = 1_A \text{ and } gf = 1_{A'}]$$

Usually the name **group** is reserved for those groupoids that are moreover monoids, but frequently in applications groupoids arise that are not monoids; fortunately, almost all the profound theorems proved in the past 200 years for one-object groups extend rather effortlessly to groups with many objects. Certainly for our purposes, the property of having all maps invertible will be more significant as an indicator of a qualitative particularity of variation/motion; the accident of being described in terms of a single object will mainly serve the subjective purpose of making some problems seem simpler. One sometimes says that a groupoid A arises concretely if we are given a category $\mathcal{E}$ of structured spaces and we define A to be the category of all isomorphisms in $\mathcal{E}$

$$A = \text{Iso}(\mathcal{E})$$

that is, the objects of A are the same as those of $\mathcal{E}$, but the maps of A are only those maps of $\mathcal{E}$ that have 2-sided inverses in $\mathcal{E}$.

Exercise 10.12
1_A is invertible for all A. If f_1, f_2 are invertible and composable in $\mathcal{E}$, then $f_2 f_1$ is invertible in $\mathcal{E}$. Hence, Iso($\mathcal{E}$) is a category.   ◊

Exercise 10.13
If g is the two-sided inverse for f, then g itself has a two-sided inverse. Hence Iso($\mathcal{E}$) is a groupoid.   ◊

Exercise 10.14
If we intersect the groupoid Iso($\mathcal{E}$) with the endomorphism monoid Endo$_\mathcal{E}(A)$ of an object, we get

$$\text{Aut}_\mathcal{E}(A) = \text{Iso}(\mathcal{E}) \cap \text{Endo}_\mathcal{E}(A)$$

a one-object groupoid (called the **automorphism group** of A in $\mathcal{E}$).   ◊

Exercise 10.15

If A is a finite set in the category $\mathcal{E} = \mathcal{S}$ of abstract sets and arbitrary maps, then

$$\mathrm{Aut}_{\mathcal{S}}(A) = A! \text{ (factorial)}$$

$\Diamond$

Exercise 10.16

Show that there is a group with only three maps

$$1, w, w^2, w^3 = 1$$

Can it be realized concretely?

$\Diamond$

10.2 Actions

10.2.1 Actions as a Typical Model of Additional Variation and the Failure of the Axiom of Choice as Typical

The notion of "action" is defined for any (even abstractly given) category $\mathcal{A}$, but then the actions of $\mathcal{A}$ are the objects of a category $\hat{\mathcal{A}}$ of structured spaces from which we can extract concrete subcategories $\mathcal{A}'$, which in turn (in their abstract guise) have actions. This dialectic is quite essential in all parts of mathematics since even if we start with the "self-action" of concrete $\mathcal{A}$, other actions of the same $\mathcal{A}$ immediately arise by mathematical constructions such as taking the spaces of open subsets, of compact subsets, of continuous functions, or of measures, on the original spaces. Actually, there are left actions and right actions of $\mathcal{A}$. Let us first consider the situation in which $\mathcal{A}$ is a given subcategory of a category $\mathcal{E}$. For any object X of $\mathcal{E}$ we can consider the algebra $\mathcal{A}(X)$ of $\mathcal{A}$-valued functions on X (in the sense of $\mathcal{E}$) as follows:

(0) For each object A of $\mathcal{A}$ there is a set

$$\mathcal{A}(X)(A) \stackrel{\text{def}}{=} \mathcal{E}(X, A)$$

defined in this case as the set of all $\mathcal{E}$-maps $X \longrightarrow A$.

(1) For each element f of the set $\mathcal{A}(X)(A)$ and for each map $A \stackrel{\alpha}{\longrightarrow} A'$ of $\mathcal{A}$ there is an associated element αf of the set $\mathcal{A}(X)(A')$, defined in this case as composition.

(2) $1_A f = f$.

(3) $(\bar{\alpha}\alpha)f = \bar{\alpha}(\alpha f)$ in $\mathcal{A}(X)(A'')$ whenever $A \stackrel{\alpha}{\longrightarrow} A' \stackrel{\bar{\alpha}}{\longrightarrow} A''$ in $\mathcal{A}$ and whenever f is an element of the set $\mathcal{A}(X)(A)$.

Any system of sets associated to the objects of a category $\mathcal{A}$ together with any system of operations between these sets associated to the maps of $\mathcal{A}$ in such a way as to satisfy conditions (2) and (3) is called a **left $\mathcal{A}$-action,** or simply a **left**

$\mathcal{A}$-**set,** regardless of whether it is given by composition in terms of some enveloping category $\mathcal{E}$ and some object X in it.

Exercise 10.17
The preceding description of the action of maps in $\mathcal{A}$ as operations on the function algebra of X is not limited to unary operations because not only is $\mathcal{A}$ not necessarily a monoid, but it even may contain Cartesian products for some of its objects. For example, if $\mathcal{E}$ is the category of smooth (C^∞) manifolds X (e.g., open balls and their bounding spheres, tori, $I\!R^n$, etc.) and smooth (C^∞) maps between them, we could take as $\mathcal{A}$ the subcategory of $\mathcal{E}$ determined by the two objects $I\!R$ and $I\!R^2$; then if f_1, f_2 are any two elements of $\mathcal{A}(X)(I\!R)$ and if $I\!R^2 \xrightarrow{\theta} I\!R$ is addition or multiplication, then there is a uniquely determined element f of $\mathcal{A}(X)(I\!R^2)$ for which $f_1\theta f_2 = \theta f$ in $\mathcal{A}(X)(I\!R)$.   ◊

The totality of left $\mathcal{A}$-sets becomes a category $\mathcal{S}^{\mathcal{A}}$ by defining the appropriate concept of map between them; in this case maps φ are often called $\mathcal{A}$-**homomorphisms** or $\mathcal{A}$-**natural maps** (or even $\mathcal{A}$-homogeneous maps). The crucial formula expressing compatibility with the action is

$$(\alpha g)\varphi = \alpha(g\varphi)$$

for all elements g of the domain of φ and all $\mathcal{A}$-maps α acting on both domain and codomain of φ; since each of the domain and codomain is really a family of sets, one for each object A of $\mathcal{A}$, correspondingly φ is really a family of set maps φ_A, and the preceding formula is often expressed as the commutativity of the naturality diagram

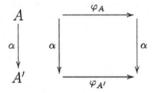

(Since left $\mathcal{A}$-sets are often "algebras" – as opposed to the "geometries" we will usually deal with – it would not be inappropriate here to use I. N. Herstein's convention that in algebra evaluation and composition are written in the opposite order: thus, $(g)\varphi$ would mean the value of the homomorphism φ at the element g, and $\varphi\psi$ would mean first φ then ψ; we will not be using that notation in most contexts.) We can sum up the discussion with the following definitions.

Definition 10.18: *If $\mathcal{A}$ and $\mathcal{B}$ are categories, a* **functor** *Φ from $\mathcal{A}$ to $\mathcal{B}$ is an assignment of*

- *an object $\Phi(A)$ in $\mathcal{B}$ for every object A in $\mathcal{A}$*
- *an arrow $\Phi(f) : \Phi(A) \longrightarrow \Phi(A')$ in $\mathcal{B}$ for every arrow $A \longrightarrow A'$ in $\mathcal{A}$*

subject to the following equations:

- $\Phi(1_A) = 1_{\Phi(A)}$
- $\Phi(gf) = \Phi(g)\Phi(f)$ *whenever* $A \xrightarrow{f} A' \xrightarrow{g} A''$

Definition 10.19: *The* **category of left** $\mathcal{A}$**-sets** $\mathcal{S}^\mathcal{A}$ *has as objects the functors* $\Phi :$ $\mathcal{A} \longrightarrow \mathcal{S}$. *A morphism* φ *in* $\mathcal{S}^\mathcal{A}$ *from* Φ *to* Φ' *is a natural transformation, where a* **natural transformation** *or* **homomorphism** $\varphi : \Phi \longrightarrow \Phi'$ *is a family of morphisms* $\varphi_A : \Phi(A) \longrightarrow \Phi'(A)$ *satisfying*

$$\Phi'(\alpha)\varphi_A = \varphi_{A'}\Phi(\alpha) \quad whenever \quad A \xrightarrow{\alpha} A'$$

Exercise 10.20
If $\mathcal{A} \subset \mathcal{E}$ is a full subcategory and if $X \xrightarrow{\varphi} Y$ is any map in $\mathcal{E}$, then there is a natural homomorphism of left $\mathcal{A}$-sets

$$\mathcal{A}(Y) \longrightarrow \mathcal{A}(X)$$

between the associated function algebras defined by

$$g \mapsto g\varphi$$

$\Diamond$

Instead of trying to maintain two conventions on the order of composition (one for algebra and one for geometry) most authors write something like

$$\mathcal{A}(Y) \xrightarrow{\varphi^*} \mathcal{A}(X)$$

to denote the operation of "pulling back along φ" the $\mathcal{A}$-valued functions g on Y to $\mathcal{A}$-valued functions on X, i.e. $\varphi^* g = g\varphi$. One can prove the functoriality formula

$$(\psi\varphi)^* = \varphi^*\psi^*$$

very easily in this context. (The functor is objectively contravariant as X varies no matter how we subjectively write it.)

Some basic examples of **right** $\mathcal{A}$**-actions** also arise from some realization of $\mathcal{A}$ as a concrete category and the choice of an object X in an enveloping category $\mathcal{E} \supset \mathcal{A}$. We will use (for the moment) $\mathcal{G}_\mathcal{A}(X)$ to denote the resulting right $\mathcal{A}$-set:

(0) For each object A of $\mathcal{A}$ there is a set

$$\mathcal{G}_\mathcal{A}(X)(A) \stackrel{\text{def}}{=} \mathcal{E}(A, X)$$

(1) For each element x of $\mathcal{G}_\mathcal{A}(X)(A)$ and for each map $A' \xrightarrow{\alpha} A$ of $\mathcal{A}$ there is an associated element $x\alpha$ of $\mathcal{G}_\mathcal{A}(X)(A')$.
(2) $\qquad\qquad\qquad\qquad x1_A = x$
(3) $\qquad\qquad\qquad x(\alpha\alpha') = (x\alpha)\alpha'$ in $\mathcal{G}_\mathcal{A}(X)(A'')$
$\qquad$ whenever $A'' \xrightarrow{\alpha'} A' \xrightarrow{\alpha} A$ in $\mathcal{A}$ and x is an element of $\mathcal{G}_\mathcal{A}(X)(A)$.

Just as we called the concrete example $A(X)$ of a left A-set a "function algebra" with the action of A called "algebraic operations" on the functions, and natural homogeneous maps "homomorphisms," so we could call $\mathcal{G}_A(X)$ the "geometry of figures" in X with the action of A expressing "incidence relations" among the figures x, x'; the natural homogeneous maps could be called "A-smooth". Note that if x is a *part* of X and if x' is a given figure in X, then there is at most one α such that $x' = x\alpha$, in which case we could just write $x' \longrightarrow x$ or $x' \in x$ or $x' \subseteq x$ to describe the incidence situation. However, in general the description of an incidence situation will involve specifically naming elements α of the set $\mathcal{G}_A(X)(x', x)$ of maps in A for which $x' = x\alpha$. This is because in general it is necessary to consider "singular" figures x (i.e., those that are not parts) to do justice to the geometry $\mathcal{G}_A(X)$ of X, as seen from A in $\mathcal{E}$. For example, A itself may be shaped as a closed interval $[0, 1]$, X may be a plane, and "A-smooth" may mean exactly "continuous". The figures in $\mathcal{G}_A(X)(A)$ are continuous curves in the plane, including many that are parts of the plane but also many loops x for which $x(0) = x(1)$ and hence x is "singular"! This latter is actually an incidence relation, for we can include in our little category A also the space A' consisting of one point. A figure of type A' in X is then just a specified point x' of the plane. There are two maps $A' \overset{\alpha_0}{\underset{\alpha_1}{\rightrightarrows}} A$ of particular interest here.

Exercise 10.21

Define α_0, α_1 so that the incidence relations in $\mathcal{G}_A(X)$

$$x\alpha_0 = x'$$
$$x\alpha_1 = x'$$

are equivalent to saying that the curve x is a loop at the point x'.   ◊

Exercise 10.22

If $A \subset \mathcal{E}$ and $X \overset{\varphi}{\longrightarrow} Y$ is an arrow of $\mathcal{E}$ show that there is an induced A-smooth map of right A-sets:

$$\mathcal{G}_A(X) \overset{\varphi}{\longrightarrow} \mathcal{G}_A(Y)$$

◊

Exercise 10.23

Let A^{op} be the **opposite category** for A, i.e., the category with the same objects as A, but with $A^{\mathrm{op}}(A, A') = A(A', A)$ and composition derived from A's. Show that a right A-set is the same thing as a *left* A^{op}-set and further that the category of right A-sets and all A-smooth maps is exactly $\mathcal{S}^{A^{\mathrm{op}}}$.   ◊

Briefly for right A-sets the action of A is contravariant, whereas the basic induced A-natural maps are covariant, (opposite to the situation for left A-sets).

Of both left and right $\mathcal{A}$-sets there is a supply of especially fundamental examples whose construction does not require imagining a concrete enveloping category $\mathcal{E}$, namely, we can consider $\mathcal{A}(X)$ and $\mathcal{G}_{\mathcal{A}}(X)$, where X is an object of $\mathcal{A}$ itself. These examples are usually referred to by names like "representable functors," "regular representation," or "self-actions by translation". Of course, if $\mathcal{A}$ is a monoid this supply of examples is limited to one, even though the categories of left $\mathcal{A}$-sets and right $\mathcal{A}$-sets are still quite vast and varied. But the representables are very useful probes in studying the general actions.

Exercise 10.24

(Yoneda's lemma) For any category $\mathcal{A}$, let C be an object of $\mathcal{A}$ and let Φ be any right $\mathcal{A}$-set, i.e. a family of sets parameterized by objects of $\mathcal{A}$ and a family of set-maps between these contravariantly parameterized by the maps of $\mathcal{A}$, satisfying the associative and unit laws

$$x1 = x$$
$$x(\alpha\alpha') = (x\alpha)\alpha'$$

Then any $\mathcal{A}$-smooth map

$$\mathcal{G}_{\mathcal{A}}(C) \xrightarrow{\varphi} \Phi$$

is actually

$$\varphi = \hat{x}$$

for a uniquely determined element x of $\Phi(C)$ (figure in Φ of type C). Here $\hat{x}(c) \stackrel{\text{def}}{=} xc$ for any $A \xrightarrow{c} C$. Further, the action in Φ can be seen as composition of $\mathcal{A}$-smooth maps: $x\alpha = \hat{x}\,\hat{\alpha}$ for $C' \xrightarrow{\alpha} C$. $\diamond$

Exercise 10.25

(Cayley) Every abstract monoid $\mathcal{A}$ can be realized concretely.

Hint: Take $\mathcal{E} = \mathcal{S}^{\mathcal{A}\text{op}}$, and take for A the (essentially unique) representable right $\mathcal{A}$-set; apply Yoneda's lemma with $\Phi = C = A$. $\diamond$

Exercise 10.26

(Dedekind, Hausdorff) Every abstract poset $\mathcal{A}$ can be realized concretely as inclusions.

Hint: Take $\mathcal{E} = \mathcal{S}^{\mathcal{A}\text{op}}$ and consider the particular object 1 such that

$$1(A) = \text{a one-element set for all } A \text{ in } \mathcal{A}.$$

Then there is a unique part

$$\mathcal{G}_A(C) \hookrightarrow 1$$

in $\mathcal{E}$ for each C in $\mathcal{A}$, and by Yoneda

$$\left[\mathcal{G}_A(B) \subseteq \mathcal{G}_A(C) \quad \begin{array}{c} \text{as parts of } 1 \\ \text{in } \mathcal{E} \end{array} \right] \Longleftrightarrow \left[\begin{array}{c} B \longrightarrow C \\ \text{in } \mathcal{A} \end{array} \right]$$

If Φ is an arbitrary right $\mathcal{A}$-set, then the action of $B \longrightarrow C$ on an element x of Φ at C is represented as composition (in $\mathcal{E} = \mathcal{S}^{\mathcal{A}^{op}}$) with the inclusion

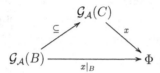

(composition of x with an inclusion is in general often called "restriction" of x from the larger part to the smaller). ◊

Exercise 10.27
The action by restriction along an inclusion is not necessarily either surjective or injective.

Hint: Consider the category $\mathcal{A} = [U \longrightarrow C]$ consisting of only two objects and only one nonidentity map. It is a poset. Let $\Phi(U) = $ the set of all continuous real-valued functions on the open interval $(0, 1)$, but $\Phi(C) = $ the set of all real functions continuous on the closed interval $[-1, 1]$. Show that the single restriction process $\Phi(C) \longrightarrow \Phi(U)$ is neither injective nor surjective. ◊

Exercise 10.28
If $\mathcal{A}$ is just a set of objects (no nonidentity maps), then

$$\mathcal{S}^{\mathcal{A}^{op}} \cong \mathcal{S}/\mathcal{A}.$$

◊

10.3 Reversible Graphs

Let $\mathcal{A}$ be the concrete four-element monoid obtained as endomaps of the two-element set in the category of sets. The concrete action of $\mathcal{A}$ on 2 is a left action, and indeed the discussion of "winning teams" in Proposition 8.22 and Exercise 8.24 concerns function algebras and homomorphisms relative to this category. However, we want here to describe instead some examples of right actions of this same $\mathcal{A}$. In this case we can visualize the right $\mathcal{A}$-sets as (reversible) "graphs" in the following way. Suppose X is a set equipped with a right action of $\mathcal{A} = \{1, \tau, c_0, c_1\}$. Since

$$c_i c_j = c_i$$

holds in $\mathcal{A}$, one can show that there is a part X_0 of the underlying set of X that can be described in several equivalent ways:

Exercise 10.29
Each of the following four conditions on an element x of X implies the others

$$x c_0 = x, \qquad x c_1 = x$$
$$\exists x'[x' c_0 = x], \qquad \exists x'[x' c_1 = x]$$

$\Diamond$

Let $X_0 \xrightarrow{i} X$ be the part of X whose members are all these special elements; then we have

$$\tau^* i = i$$

where τ^* is the map defined by the action

$$\tau^* x = x\tau$$

Moreover, there exists a unique pair of maps d_0, d_1 as in

$$X_0 \underset{d_1}{\overset{i}{\underset{\longrightarrow}{\overset{d_0}{\longleftarrow}}}} X \circlearrowright \tau^*$$

such that $i d_0 = c_0{}^*, i d_1 = c_1{}^*$, where $c_k{}^* x = x c_k$ as with τ. Then

$$d_0 \tau^* = d_1, \qquad d_1 \tau^* = d_0$$

The elements of X_0 are often called "nodes" or "vertices".

Definition 10.30: *For $\mathcal{A} = \{1, \tau, c_0, c_1\}$, an element x of a right $\mathcal{A}$-set is called a* **loop** *if and only if*

$$x c_0 = x c_1$$

An element x is called a **one-lane loop** *if and only if*

$$x\tau = x$$

Exercise 10.31

If x is a loop, then $x\tau$ is also a loop. Any one-lane loop is a loop (but not conversely), and any member of X_0 is a one-lane loop (but not conversely).   ◊

The preceding calculations suggest a good way to picture an $\mathcal{A}$-set X:

Definition 10.32: *Say that a* $\xrightarrow{\ x\ }$ *b in X if and only if*

$$xc_0 = a \text{ and } xc_1 = b$$

(*Caution*: X is not usually a category since we are not given any composition rule, and hence x is not in itself a map since we cannot apply it to anything – on the other hand, we will sometimes consider various ways of enlarging X to get a category.)

Thus, we can picture any right $\mathcal{A}$-set as a reversible graph. The general elements of X are pictured as arrows that connect definite nodes as specified by the action of c_0, c_1. For a node a it is convenient to picture the one-lane loop ia as an arrow that is collapsed to a point, whereas all other arrows (including other loops and even other one-lane loops) retain length. Since

$$a \xrightarrow{\ x\ } b \iff b \xrightarrow{\ x\tau\ } a$$

it is convenient to draw x, $x\tau$ next to one another like the two lanes of a two-lane highway. Thus,

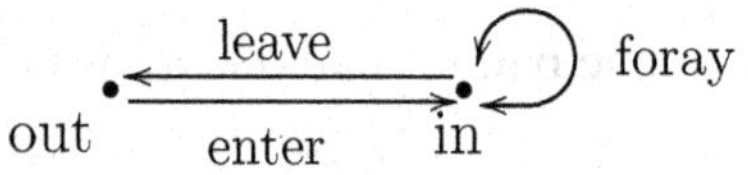

is a very important example of a five-element right $\mathcal{A}$-set that has no two-lane loops. [The representable right $\mathcal{A}$-set, the self-action, gets instead the following picture

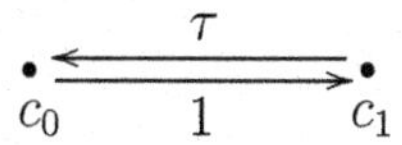

The five-element example will be seen to consist of the "truth values" for the whole category $\mathcal{S}^{\mathcal{A}^{op}}$.

Exercise 10.33

There is a unique $\mathcal{A}$-smooth (i.e., natural homogeneous) map φ from the representable right $\mathcal{A}$-set into the truth-value graph for which

$$\varphi(1) = \text{enter}$$

There is another one ψ uniquely determined by

$$\psi(1) = \text{foray}$$

φ is injective and hence a part in $\mathcal{S}^{\mathcal{A}^{\mathrm{op}}}$ but ψ is not injective on either nodes or arrows. ◊

Exercise 10.34
Define (by a table) an action of $\mathcal{A}$ on a set of the right size so that the resulting $\mathcal{A}$-set can be pictured as the graph

and do the same for

◊

Exercise 10.35
Define a surjective $\mathcal{A}$-smooth map p between the two $\mathcal{A}$-sets above

Determine all $\mathcal{A}$-smooth maps s in the opposite direction and show that

$$ps \neq id$$

for all of them. ◊

Exercise 10.36
Define a surjective $\mathcal{A}$-smooth map p "from the interval to the loop"

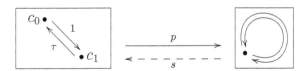

Determine all $\mathcal{A}$-smooth s in the opposite direction and show that $ps \neq id$ for all such s. ◊

The last two exercises illustrate that the failure of the axiom of choice is typical for variable or cohesive sets.

10.4 Chaotic Graphs

Several different kinds of structures are loosely referred to as "graphs"; for example, we have already briefly introduced "reversible" graphs. In this section (and frequently later) we will call "graphs" the reflexive, not necessarily reversible graphs, which are just the right actions of the three-element monoid

$$\Delta_1 = \{1, \delta_0, \delta_1\} \quad \delta_i\delta_j = \delta_i \quad i, j = 0, 1$$

Exercise 10.37

This monoid can be realized concretely as the order-preserving self-maps of the nontrivial two element poset.   ◊

If X is such a graph, then for each element x, $x\delta_0$ is the beginning point of the arrow x, and $x\delta_1$ is the endpoint. The points themselves are considered degenerate arrows, which are special loops.

Definition 10.38: *The arrow x is a* **loop** *if and only if $x\delta_0 = x\delta_1$ and is* **degenerate** *if and only if*

$$x\delta_0 = x = x\delta_1$$

Exercise 10.39

If $x\delta_0 = x$, then x is a degenerate loop (use the multiplication table for Δ_1 and the associativity of the action).   ◊

Exercise 10.40

Write the action table for the seven-element right Δ_1-set pictured here

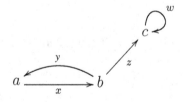

                                                                    ◊

A basic example of a graph is the right action of Δ_1 on itself, which is pictured as

$$0 \xrightarrow{\ u\ } 1$$

with the correspondences

$$u \sim 1, \quad 0 \sim \delta_0, \quad 1 \sim \delta_1 \text{ in } \Delta_1$$

Exercise 10.41

If X is any graph and x in X, then there is a unique map of graphs (= Δ_1-smooth map)

for which $\bar{x}(u) = x$. ◊

We now want to discuss *left-Δ_1* actions. There is a huge difference between the left and right categories. The graphs have the capacity to encode arbitrarily complicated information, and thus the complete classification of graphs is probably not possible. On the other hand the *left-actions* can be completely classified almost immediately. Their interest is that they will give rise to a class of *right* actions, the chaotic graphs, which will include the recipients for many graph quantities of interest.

In general, if T is a set with a given *left* action of a monoid $\mathcal{A}$, and if V is any set, then the set

$$T \Rightarrow V$$

of all maps from T to V has a natural *right* action of $\mathcal{A}$:

$$(f\alpha)(t) = f(\alpha t) \text{ for all } t \text{ in } T$$

which defines $f\alpha$ as a new element of $T \Rightarrow V$ for any given element f of $T \Rightarrow V$. The argument αt (at which f is evaluated to get the value at t of $f\alpha$) is given by the presupposed left action on T of α in $\mathcal{A}$.

Exercise 10.42

For an iterated action $f(\alpha_1\alpha_2) = (f\alpha_1)\alpha_2$ since both sides have the same value at each t in T. ◊

Now in particular a left action on T of the three-element monoid Δ_1 means

$$\delta_i\delta_j t = \delta_i t \quad \text{for } i, j = 0, 1$$

Exercise 10.43

If $\delta_0 t = \delta_0 s$, then $\delta_1 t = \delta_1 s$ and conversely. ◊

Hence, the orbits of a left action are just determined by either of the two equations in the exercise, and each of δ_0, δ_1 is constant *on each orbit*. In other words, a left action with just one orbit is specified by an arbitrary set and a chosen pair t_0, t_1 of elements, with

$$\delta_0 t = t_0$$
$$\delta_1 t = t_1$$

for all t in the orbit.

(Recall that the basic example of a left Δ_1 action involved two *constant* maps, but the equations cannot force that to be taking place only on a two-element set.) The most general left Δ_1-set is just a disjoint sum (no interaction between the summands) of summands, each of which is a left Δ_1 set in its own right but with the special property of having a single orbit. Since

$$\left(\sum_k T_k\right) \Rightarrow V \cong \prod_k (T_k \Rightarrow V)$$

is a general result, for our present purpose it will suffice to consider the case of a single orbit. However,

Exercise 10.44

If $\pi_0(T)$ denotes the set of orbits (components) of a left Δ_1-set T, then the action of δ_0, δ_1 gives rise to two sections of the canonical surjection $T \longrightarrow \pi_0(T)$:

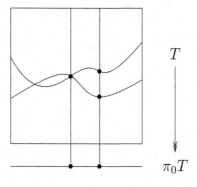

$\diamondsuit$

The orbits are the fibers; each of these contains two labeled points t_0, t_1, but in some fibers the labeled points may coincide. An arbitrary left Δ_1-set is described up to isomorphism by giving the set I of components ($I = \pi_0(T)$ above) for each i in I, giving the number N_i of elements in that fiber, and specifying the part I_1 of the components in which $t_0 = t_1$ (rather than $t_0 \neq t_1$) is to hold; these are restricted only by $I_1 \leq I$ and

$$N_i \geq 1 \text{ for all } i \text{ in } I$$
$$N_i \geq 2 \text{ for all } i \text{ not in } I_1$$

(I itself can be 0.)

Returning to the construction of the chaotic graph $T \Rightarrow V$ with a given set V of "values" and a given left Δ_1-set T with one component: the arrows are arbitrary functions f from T to V, and

$$(f\delta_0)(t) = f(\delta_0 t) = f(t_0)$$
$$(f\delta_1)(t) = f(\delta_1 t) = f(t_1)$$

define $f\delta_0$ and $f\delta_1$ as constant functions.

Exercise 10.45
The points of $T \Rightarrow V$ are all the constant functions. Identifying these points with the elements of V, the starting and ending points of any f are identified with the values at t_0 and t_1 of f, respectively. $\diamond$

The intuitive reason for calling $T \Rightarrow V$ "chaotic" is that, for given points v_0, v_1, *any* function $T \xrightarrow{f} V$ for which $f(t_0) = v_0$ and $f(t_1) = v_1$ is considered admissible as an arrow from v_0 to v_1. One way of relating these very special graphs to more general, more highly structured ones, is indicated as follows:

Exercise 10.46
If X is any graph and if R is the chaotic graph $R = T \Rightarrow V$ derived from the value set V and the left Δ_1-set of one component given by $1 \underset{t_1}{\overset{t_0}{\rightrightarrows}} T$, then a graph map $X \xrightarrow{f} R$ is determined by specifying a value $f(x)$ in V for each point of V and a function $f_x : T \longrightarrow V$ for each arrow x of X subject to these conditions:

(1) if $a \xrightarrow{x} b$ in X then

$$t = t_0 \Rightarrow f_x(t) = f(a)$$
$$t = t_1 \Rightarrow f_x(t) = f(b)$$

(2) If $x = x\delta_0 = x\delta_1 = a$ is a degenerate arrow of X, then f_x is the constant function $f_x(t) = f(a)$ for all t in T. $\diamond$

Among the chaotic graphs $T \Rightarrow V$ are the codiscrete ones determined by an arbitrary set V and by the extremely special two-element left Δ_1-set T in which $t_0 \neq t_1$. In other words the codiscrete graph on V points has as its arrows all ordered pairs of elements of V with

$$\langle v_0, v_1 \rangle \delta_i = \langle v_i, v_i \rangle \quad \text{for } i = 0, 1$$

That is, the projection and diagonal structure on the cartesian product $V \times V$ determines the codiscrete graph structure. The universal mapping property becomes simply the following:

Proposition 10.47: *If $\underline{V}$ is the codiscrete graph with V points and if X is any graph, then any graph map*

$$X \longrightarrow \underline{V}$$

is entirely determined by a map $X_0 \longrightarrow V$ of sets, where X_0 is the set of points of X. That is, the restriction

$$\left(X \Rightarrow_{\Delta_1^{op}} \underline{V} \right) \longrightarrow (X_0 \Rightarrow_1 V)$$

is an isomorphism of abstract sets.

Proof: No matter how many arrows $a \xrightarrow{x} b$ there are in X, $f(x)$ can only be the arrow $\langle f(a), f(b) \rangle$ in $\underline{V}$, independently of x.   ■

Corollary 10.48: *There is a graph $\underline{2}$ such that for any graph X, the graph maps $X \longrightarrow \underline{2}$ are in one-to-one correspondence with the arbitrary subsets of the set of points (vertices) of X.*

Proof: The set 2 of two elements is the truth-value object for the category of abstract sets. So consider the codiscrete graph $\underline{2}$ with four arrows and two points, and apply Proposition 10.47.   ■

One can similarly "classify" sets of arrows (not just of vertices).

There is a completely different two-element left Δ_1-set T, namely, the one in which $\delta_0 t = \delta_1 t$ for all t; in other words $t_0 = t_1$, but there is one other element t that is not a value of either δ_0 or δ_1.

Exercise 10.49
For the left Δ_1-set T just described, the graph $T \Rightarrow V$, for any given set V, satisfies

$$T \Rightarrow V \cong V \times \dot{V}$$

the sum of V noninteracting copies of a graph $\dot{V}$ that has one point and V loops. (Under the isomorphism the "name" of the degenerate loop depends on the summand.)   ◊

Note that a "pointed set" can be viewed as either a graph or a left Δ_1-set. This can be explained as follows:

Exercise 10.50

If $\mathcal{A}$ is a commutative monoid, there exists an isomorphism of categories

$$\text{left } \mathcal{A}\text{-sets} \longrightarrow \text{right } \mathcal{A}\text{-sets}$$

Note: An **isomorphism of categories** is a functor that has an inverse, which means a functor in the opposite direction so that both composites of the two functors are identity functors. ◇

Exercise 10.51

There is a commutative monoid $\{1, \epsilon\}$ with two elements in which $\epsilon^2 = \epsilon$ and a surjective homomorphism (= functor) of monoids

$$\Delta_1 \longrightarrow \{1, \epsilon\}$$

Hence, any left (respectively right) $\{1, \epsilon\}$-set has (via the homomorphism) a natural left (respectively right) Δ_1 action. ◇

Exercise 10.52

In the notation of the preceding exercise, any $\{1, \epsilon\}$-set is uniquely a disjoint sum of pointed sets. ◇

Exercise 10.53

A left Δ_1-set arises by the process above from a pointed set if and only if it has a single orbit (of arbitrary size T), and in that orbit the two distinguished points are equal: $t_0 = t_1$. ◇

Exercise 10.54

A right Δ_1-set arises by the procedure above from a pointed set if and only if it is a graph with precisely one point (= vertex), i.e. if and only if it is connected and all arrows are loops. ◇

Exercise 10.55

A graph is of the form $T \Rightarrow V$, where T is a left Δ_1-set arising from a pointed set (as above) if and only if it is a graph consisting entirely of loops and in which every vertex is the site of the same number of loops, which number is moreover a power of the total number of vertices. More exactly, if $T = E + \{t_0 = t_1\}$, then one can construct an isomorphism of graphs

$$V \times (\dot{V}^E) \xrightarrow{\sim} T \Rightarrow \dot{V}$$

◇

10.5 Feedback and Control

Can a graph $\langle 1_A, \varphi \rangle$ of a map $A \xrightarrow{\varphi} B$ be part of a graph structure in the sense of a right Δ_1-set? One way of making this question definite is as follows: A right Δ_1-set P can equivalently be described as a pair of sets and three maps

$$A \overset{\displaystyle \underset{\psi}{\overset{\pi}{\underset{\gamma}{\longleftrightarrow}}}}{\longrightarrow} P$$

for which $\pi \gamma = 1_A = \psi \gamma$; that is, the "reflexivity" γ is a common section for the "source" map π and for the "target" map ψ so that it chooses for each "state" a in A a particular "arrow" γa in P, which is actually a loop, i.e. an arrow from a to a itself. The right Δ_1-set structure on such a P is defined by $p \cdot \delta_0 = \gamma \pi p$, $p \cdot \delta_1 = \gamma \psi p$.

Exercise 10.56

Verify that these equations define a right Δ_1-set. Conversely, show that any right Δ_1-set P gives rise to such a diagram by defining the state set A to consist of the trivial arrows, i.e. those p for which $p \cdot \delta_0 = p = p \cdot \delta_1$. Remember that not all loops are trivial. $\Diamond$

Exercise 10.57

If $A \xrightarrow{\varphi} B$ and $\gamma = \langle 1_A, \varphi \rangle$ is its graph, then on the set $P = A \times B$ we have part of a graph structure

$$A \overset{\displaystyle \underset{\psi}{\overset{\pi}{\underset{\gamma}{\longleftrightarrow}}}}{\longrightarrow} A \times B$$

where $\pi = $ projection to A since $\pi \gamma = 1_A$. (The map φ is recovered from its graph γ by following instead with the projection to B.) Show that a map $A \times B \xrightarrow{\psi} A$ completes the structure of a right Δ_1-set on $A \times B$ iff $\psi(a, \varphi a) = a$ for all a. $\Diamond$

Given any $A \times B \xrightarrow{\psi} A$ to be considered as "target," pairing it with the projection π to A considered as "source," we get a "graph" (not reflexive) in which the arrows are labeled by the elements of B with the property that given any b and any state a there is exactly one arrow in the graph with label b and source a; the target of that arrow is $a' = \psi(a, b)$. The elements of B act via ψ on the states, and the graph pictures the possible operation of the system; imagine that if b is activated when the system is in state a, the system will move somehow to state a'. The element b may be said to maintain a state a iff $a = \psi(a, b)$, i.e. if b labels a loop at a. The problem of completing the information ψ with a map φ whose graph $\gamma = \langle 1, \varphi \rangle$

makes this graph reflexive may be a nontrivial one if the maps are to be in a highly structured category such as a category from linear algebra.

Exercise 10.58

Let A and B be vector spaces and let $B \xrightarrow{\alpha} A$ be a linear map. We will consider some action graphs in which α is one of the ingredients. A simple example is essentially just the vector subtraction in A

$$\psi(a, b) = \alpha(b) - a$$

Show that, with this action graph, b maintains a iff

$$\alpha(b) = 2a$$

$\Diamond$

Exercise 10.59

Using a "feedback" map $A \xrightarrow{\beta} B$ to "observe" the states permits one to adjust the acting element b before applying α, leading to an action called **feedback control**

$$\psi(a, b) = \alpha(b - \beta(a))$$

Show that b maintains a with feedback control iff

$$\alpha(b) = (1 + \alpha\beta)a$$

$\Diamond$

Exercise 10.60

Given two opposed linear maps, the graph defined by feedback control ψ as above may sometimes admit a completion to a right Δ_1 action on $A \times B$; this requires another linear map $A \xrightarrow{\varphi} B$ whose graph would supply the reflexivity; in other words, φ would supply, for every state a, a chosen b that would succeed in maintaining a. Show that a map φ works that way iff

$$\alpha\varphi = 1 + \alpha\beta$$

Show that if solutions φ to the latter equation exist, then α is surjective (and one can take $\varphi = \sigma + \beta$, where σ is any chosen section of α). But note that in most practical examples, $\dim A > \dim B$, so that no surjective α can exist. $\Diamond$

The preceding exercise raises the question of calculating the possible reflexivity structures on subgraphs of a given nonreflexive graph $A \overset{\pi}{\underset{\psi}{\rightleftarrows}} P$. Thus, we consider homomorphisms

$$E \xrightarrow{\ h_P\ } P$$

where $\sigma\rho = 1_V = \tau\rho, \quad \pi h_P = h_A\sigma, \quad \psi h_P = h_A\tau.$

Exercise 10.61

Show how to consider the diagram above as $q^*(Y) \xrightarrow{\ h\ } X$, where X is the given $A \underset{\psi}{\overset{\pi}{\leftleftarrows}} P$ and where q is the inclusion functor between finite categories

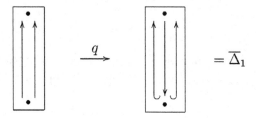

where $\overline{\Delta}_1$ is a category with seven maps all told, the bottom object is in two ways a retract of the top object, and the endomorphisms of the top are isomorphic to Δ_1.   ◊

Exercise 10.62

For every X as above there is a well-defined reflexive graph (right Δ_1-set) $q_*(X)$ such that for every Y there is a bijection

$$\frac{q^*Y \longrightarrow X}{Y \longrightarrow q_*X}$$

between (nonreflexive) graph maps above the bar and Δ_1-maps (i.e., graph maps preserving given reflexivities) below the bar. In particular, each way of choosing a reflexivity structure Y on a part of X can be indicated by factoring across the canonical $q^*q_*X \longrightarrow X$.   ◊

Returning to the linear construction discussed in Exercises 10.51 through 10.59, we point out that it also is a functor. That is, we can consider diagrams

$$A \xrightarrow{\ \beta\ } B \qquad A' \xrightarrow{\ \beta'\ } B'$$

of linear maps as concrete structures of a certain abstract kind (see Exercise 10.63).

Exercise 10.63
Define morphisms between such structures in order to obtain a reasonable category $\mathcal{D}$ of feedback-command data.

Hint: A morphism should consist of a pair of linear maps satisfying two commutativity equations. ◇

Exercise 10.64
Show that there is a functor $\mathcal{F}$ from $\mathcal{D}$ to the category $Aff^{\downarrow\downarrow}$ of nonreflexive affine graphs such that

$$\mathcal{F}(A \underset{\alpha}{\overset{\beta}{\rightrightarrows}} B) = A \underset{\psi}{\overset{\pi_A}{\leftleftarrows}} A \times B$$

with $\psi(a, b) = \alpha(b) - a$ as before. ◇

Exercise 10.65
Given any diagram X of shape $A \underset{\alpha}{\overset{\beta}{\rightrightarrows}} B$ in any category, consider the diagram $B(X)$ of the same shape

$$B \underset{\beta\alpha}{\overset{1_B}{\rightleftarrows}} B$$

Show that there is a natural map $X \overset{\bar{\bar{\beta}}}{\longrightarrow} B(X)$ of such diagrams (one leg of which is 1_B). In the linear case, determine the graph $\mathcal{F}(B(X))$ and explain why the functorially induced $\mathcal{F}(\bar{\bar{\beta}})$ "labels" the arrows in the graph $\mathcal{F}(X)$. ◇

Exercise 10.66
Linearize and functorize Exercises 10.61 and 10.62. ◇

10.6 To and from Idempotents

Some constructions starting from an arbitrary given map may produce diagrams involving idempotent maps $e = e \cdot e = e^2$. In most categories idempotents can be split, giving structures of the abstract shape

wherein $pi = 1$, $e = ip$. Note that the category $\mathbf{2} = \boxed{\cdot \longrightarrow \cdot}$ has the property that functors $\mathbf{2} \longrightarrow \mathbf{A}$ to any category $\mathbf{A}$ correspond exactly to morphisms in $\mathbf{A}$.

Exercise 10.67

Consider the three functors

$$2 \underset{\underset{e}{\overset{i}{\Longrightarrow}}}{\overset{p}{\Longrightarrow}} \mathbf{E}$$

corresponding to the morphisms of the same names in **E**. Given any σ in S with $\sigma^2 = \sigma$, each of $p^*(\sigma), i^*(\sigma), e^*(\sigma) = \sigma$ is a map in S considered as a left **2**-set. Given any $X \overset{f}{\longrightarrow} Y$ in S considered as a left **2**-set M, there are thus six sets equipped with idempotents that are universal with respect to maps to or from M. Show that these adjoints give specifically

$p!M =$ cograph of f
$p_*M =$ kernel pair (equivalence relation) of f
$i!M =$ cokernel pair of f, gluing two copies of Y along X
$i_*M =$ graph of f

all equipped with a natural idempotent, whereas

$e!M =$ cograph of f but with different adjunction map than $p!M$
$e_*M =$ graph of f but with different adjunction map than i_*M.

Determine also the other two associated canonical adjunction maps.   ◊

Exercise 10.68

Splitting the idempotents in a left action of Δ_1 yields a diagram of the form of a left action of $\overline{\Delta_1}$,

$$T \underset{\underset{s_1}{\overset{p}{\longleftarrow}}}{\overset{s_0}{\longrightarrow}} X$$

with the two s_0, s_1 having a common retraction p. It is sometimes said that p "unites the two identical opposites" since the two subobjects s_0, s_1 united in the one X are identical (with T) as mere objects (i.e., upon neglecting the inclusions) but are often "opposite" in one sense or another. Because the two are subobjects of the same object, we can intersect them, obtaining the subobject $s_0 \cap s_1$ of X. Since they also have a common domain, we can consider their equalizer k. However, the presence of p implies that these two constructions give the same result. That is, the natural map f, which proves inclusion in one direction,

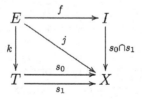

exists because $s_0 t = s_1 t = x$ implies that $x \in s_0$ and $x \in s_1$, but using p we can also show that for any x in X, if

$$t_0 \text{ proves } x \in s_0 \text{ and } t_1 \text{ proves } x \in s_1$$

then in fact $t_0 = t_1$ so that actually $x \in j$, thus defining an inverse for f. ◊

10.7 Additional Exercises

Exercise 10.69
In a poset $\mathcal{A}$ show that $A \leq A'$ and $A' \leq A$ implies $A \cong A'$.

Exercise 10.70
From a poset $\mathcal{A}$ may be constructed a *strict* poset $\mathcal{A}_s$ as follows. There is an equivalence relation R on objects of $\mathcal{A}$ defined by $A R A'$ iff $A \leq A'$ and $A' \leq A$. $\mathcal{A}_s$ has objects given by the codomain $\mathcal{A}_s$ of the associated partition $\mathcal{A} \xrightarrow{p_R} \mathcal{A}_s$. There is an arrow $B \leq B'$ in $\mathcal{A}_s$ iff there are $A \leq A'$ in $\mathcal{A}$ such that $p_R(A) = B$ and $p_R(A') = B'$.

(a) Show that $\mathcal{A}_s$ is a strict poset and that p_R is a functor (order-preserving).

(b) Show that $\mathcal{A}_s$ satisfies a universal property: If $\mathcal{A} \xrightarrow{F} \mathcal{B}$ is a functor to a strict poset $\mathcal{B}$, then there is a unique functor $\mathcal{A}_s \xrightarrow{F'} \mathcal{B}$ such that $F' p_R = F$.

Exercise 10.71
Show that $\mathcal{S}/X$ is the same as the category of left X-sets, where X is viewed as a category with no nonidentity arrows. Compare Exercise 10.28.

Exercise 10.72
Show that the category $\mathcal{S}^{\mathcal{A}}$ has finite limits and colimits. These are all computed "pointwise". For example, if Φ and Ψ are left $\mathcal{A}$-sets, then their product in $\mathcal{S}^{\mathcal{A}}$ is a left $\mathcal{A}$-set $\Phi \times \Psi$ with $(\Phi \times \Psi)(A) = \Phi(A) \times \Psi(A)$.

Exercise 10.73
Show that the category $\mathcal{S}^{\mathcal{A}}$ has a natural number object.

Hint: All of the $\mathcal{A}$-actions in that object are identity.

Exercise 10.74
Show that the category $\mathcal{S}^{\mathcal{A}}$ has a truth value object Ω. This is a generalization of Exercise 6.15.

Hint: As a functor $\Omega : \mathcal{A} \longrightarrow \mathcal{S}$, the value of Ω at X in $\mathcal{A}$ is given by the set of "subfunctors" of the representable functor $\mathcal{G}_A(X)$.

Exercise 10.75

Show that the category $\mathcal{S}^A$ has mapping sets.

Hint: If Φ and Ψ are left $\mathcal{A}$-sets, then the functor $\Psi^\Phi : \mathcal{A} \longrightarrow \mathcal{S}$ has value at X given by the set of natural transformations from $\Phi \times \mathcal{G}_A(X)$ to Ψ. Write the precise formula for the $\mathcal{A}$-action on the latter sets.

Appendix A

Logic as the Algebra of Parts

A.0 Why Study Logic?

The term "logic" has always had two meanings – a broader one and a narrower one:

(1) All the general laws about the movement of human thinking should ultimately be made explicit so that thinking can be a reliable instrument, but

(2) already Aristotle realized that one must start on that vast program with a more sharply defined subcase.

The achievements of this subprogram include the recognition of the necessity of making explicit

(a) a limited universe of discourse, as well as

(b) the correspondence assigning, to each adjective that is meaningful over a whole universe, the *part* of that universe where the adjective applies. This correspondence necessarily involves

(c) an attendant homomorphic relation between connectives (like *and* and *or*) that apply to the adjectives and corresponding operations (like *intersection* and *union*) that apply to the parts "named" by the adjectives.

When thinking is temporarily limited to only one universe, the universe as such need not be mentioned; however, thinking actually involves relationships between several universes. For example, the three universes (1) of differential equations, (2) of functions of time, and (3) of formal power series are all distinct with different classes of adjectives meaningful over each one. But there are key relationships between these three universes that are the everyday preoccupation of users of many mathematical sciences: a function might *satisfy* a differential equation, a power series might *approximate* a function or might *formally satisfy* a differential equation, whereas we might seek to *solve* a differential equation or *expand* a function, and so on. Each suitable passage from one universe of discourse to another induces

(0) an operation of substitution in the inverse direction, applying to the adjectives meaningful over the second universe and yielding new adjectives meaningful over the first universe.

The same passage also induces two operations in the forward direction:

(1) one operation corresponds to the idea of the direct image of a part but is called "existential quantification" as it applies to the adjectives that name the parts;
(2) the other forward operation is called "universal quantification" on the adjectives and corresponds to a different geometrical operation on the parts of the first universe.

It is the study of the resulting algebra of parts of a universe of discourse and of these three transformations of parts between universes that we sometimes call "logic in the narrow sense". Presentations of algebraic structures for the purpose of calculation are always needed, but it is a serious mistake to confuse the arbitrary formulations of such presentations with the objective structure itself or to arbitrarily enshrine one choice of presentation as *the* notion of logical theory, thereby obscuring even the existence of the invariant mathematical content. In the long run it is best to try to bring the form of the subjective presentation paradigm as much as possible into harmony with the objective content of the objects to be presented; with the help of the categorical method we will be able to approach that goal.

How much logic is actually relevant to the practice of a mathematical science is a question we can only scientifically answer after learning a significant fragment of this logic. Some use of logic is essential in clarifying the successive states of our calculations in the everyday algebra of ordinary variable quantities. In this appendix we will become familiar with some instructive examples of such use after making explicit the main features of this logical algebra, both objectively in terms of parts of universes of discourse as well as subjectively in terms of the syntax and symbolism commonly used to present the logical algebra that organizes the naming of these parts. As is customary, we begin with the propositional logic of a single universe of discourse and then proceed to the predicate logic where that universe is varied; the variation of universes introduces new features even in those cases in which it only involves passing from a given universe to a universe of ordered pairs or ordered triples.

Note: In the foregoing preliminary remarks we have used the term "adjective," whereas in the remaining text of this appendix we refer usually to "statements"; the link between the two is expressed by the traditional term "predicate" as follows: Over a given universe of discourse (such as functions) an adjective (or more generally an adjectival phrase) such as "positive" corresponds to a predicate or statement form such as "is positive," which can in turn become many particular statements such as

$$1 + x^2 \text{ is positive}$$
$$e^x \text{ is positive}$$

and so on; each thing or family of things (in the universe of discourse) for which the resulting statement is true belongs to the part of the universe (e.g., the positive functions) the adjective specifies.

Historical Note

It is sometimes objected that logic is allegedly not algebra since – for example – nobody thinks in cylindric algebra. That is an unfortunate misunderstanding: cylindric algebras (and polyadic algebras) were an important initial attempt in the 1950s to make explicit the objective algebra briefly described above; however, they were excessively influenced by the styles of subjective *presentation* (of logical content) that had become traditional since the 1930s under the name of "first order predicate calculus" or the like. Those styles of presentation involved various arbitrary choices (such as the specification of a single infinite list of variables) that proved to be quite bizarre when confronted with ordinary mathematical practice; surely the logic of mathematics is not cylindric algebra. For about a hundred years now, mathematical scientists have indeed had an intuitive distrust of attempts by some logicians to interfere with mathematics. However, some explicit recognition of the role of logical algebra is helpful and even necessary for the smooth and correct learning, development, and use of mathematics, including even the high school algebra of ordinary quantities.

A.1 Basic Operators and Their Rules of Inference

Although the basic statements of mathematics are equations, many of the statements of interest have a somewhat more complicated structure; often this more complicated structure is expressed in the words that surround the equations, but in order to bring this structure into sharper relief it is useful to introduce a few "logical" symbols beyond just $=$ (equals). These are

$$\exists, \wedge, \vee, \text{ true, false } \text{(positive)}$$
$$\forall, \Rightarrow \qquad \text{(negative)}$$

used as operations to build up more complicated statements from simpler ones (usually starting from equations as the "simplest," although of course equations themselves can be complicated) and the symbol $\vdash$, which expresses that one statement *follows from* another. (The reason for distinguishing the negative or higher operators from the positive ones is mentioned in Exercise A.16.) Thus if A, B are statements, then

$$A \wedge B$$

is the statement whose truth would mean that A is true *and* B is true, whereas

$$A \vee B$$

is the statement whose truth would mean roughly that at least one of A *or* B is true; thus, $\wedge$, $\vee$ are read simply "and," "or". If we express by

$$C \vdash D$$

the relation that "D follows from C" (often called "C entails D"), then the essential *rules of inference* that govern the use of $\wedge$ are as follows:

If $C \vdash D_1$ and if $C \vdash D_2$, then

$$C \vdash D_1 \wedge D_2$$

and conversely, if $C \vdash D_1 \wedge D_2$, then both

$$C \vdash D_1 \quad \text{and} \quad C \vdash D_2$$

This can be expressed more briefly as

$$C \vdash D_1 \wedge D_2 \quad \text{iff} \quad C \vdash D_1, C \vdash D_2$$

or still more briefly as

$$\frac{C \vdash D_1 \wedge D_2}{C \vdash D_1, \quad C \vdash D_2}$$

where the horizontal bar means "iff" in this context, i.e. if we know of three statements C, D_1, D_2 that they satisfy the deducibility (or "following from") relation above the line, then we can conclude that they satisfy the deducibility relations below the line, and conversely. Since

$$C \vdash C$$

holds for any statement, we can conclude, taking the special case where C *is* $D_1 \wedge D_2$, that

$$D_1 \wedge D_2 \vdash D_1$$
$$D_1 \wedge D_2 \vdash D_2$$

for any two statements D_1, D_2, and that in turn implies the "top to bottom" half of the basic rule of inference above for $\wedge$ if we use the fact that "following from" is *transitive*:

Whenever $C \vdash E$, $E \vdash D$, then also $C \vdash D$.

To analyze the "bottom to top" half of the basic two-way rule of inference for $\wedge$ similarly, note that

(a) (when we take the case in which D_1, D_2 are both C),

$$C \vdash C \wedge C$$

for any C, and also that

(b) (exercise!) whenever $C_1 \vdash D_1$, $C_2 \vdash D_2$ it follows that

$$C_1 \wedge C_2 \vdash D_1 \wedge D_2$$

Taking the case of (b), where C_1, C_2 are both C, and applying the transitivity of $\vdash$ to that together with (a), we obtain that

whenever $\quad C \vdash D_1, C \vdash D_2, \quad$ then also $\quad C \vdash D_1 \wedge D_2$

Since we already understood what $\wedge$ means, the detailed analysis above of its characterization in terms of $\vdash$ may seem pedantic; however, it is necessary to grasp every detail of it because we will apply *exactly the same* pattern of analysis to many other concepts that will not be quite so obvious.

The statement "true" is roughly like the quantity 1. It is characterized by the fact that

$$C \vdash \text{true}$$

holds for any statement C.

Exercise A.1

The statements

$$C \vdash C \wedge \text{true} \qquad C \wedge \text{true} \vdash C$$

hold for any C. ◊

We often write $C \equiv C \wedge \text{true}$. Here $A \equiv B$ (read "A is equivalent" to B) means, as in this case, that entailment $\vdash$ holds in both directions.

To say that a statement D is true means that it follows from true:

$$\text{true} \vdash D$$

Exercise A.2

Whenever

$$\text{true} \vdash D_1$$
$$\text{true} \vdash D_2$$

then

$$\text{true} \vdash D_1 \wedge D_2$$

◊

The relation of "false" to $\vdash$ is essentially opposed to that of "true". That is,

$$\text{false} \vdash D$$

holds for *any* statement D, and

$$C \vdash \text{false}$$

means that C is false; for example $0 = 1 \vdash$ false holds in any nondegenerate ring.

The basic rule of inference for $\vee$ is opposite to that for $\wedge$:

$$\frac{C_1 \vee C_2 \vdash D}{C_1 \vdash D, \quad C_2 \vdash D}$$

Note that the comma below the line still means "and" in the metalanguage even though what we are describing is "or".

Exercise A.3

Verify that the above rule of inference is intuitively reasonable; derive from it that

$$C_i \vdash C_1 \vee C_2 \qquad i = 1, 2$$
$$D \vee D \vdash D$$

holds for any C_1, C_2, D and also that whenever $C_1 \vdash D_1, C_2 \vdash D_2$,

$$C_1 \vee C_2 \vdash D_1 \vee D_2$$

then use transitivity and the identity inference to rederive the two-way rule from these special consequences, i.e. analyze the $\vee$ rule in the "same pattern" by which we analyzed the $\wedge$ rule. $\Diamond$

Exercise A.4

(false $\vee C$) $\equiv C$, for all C $\Diamond$

The operation $\Rightarrow$ applied to a pair of statements B, D gives another statement $B \Rightarrow D$, which is usually read "B implies D" or "if B then D". (It is to be *distinguished from $B \vdash D$*, which is a statement *about* statements usually written down only when we mean to assert that D in fact follows from B, whereas $B \Rightarrow D$ is a statement which, like other statements, may be important to consider even when it is only partly true.) The rule of inference characterizing $\Rightarrow$ is not only in terms of $\vdash$, but also involves $\wedge$:

$$\frac{C \vdash (B \Rightarrow D)}{C \wedge B \vdash D}$$

(This is sometimes called "modus ponens and the deduction theorem" by logicians.) That is, if we know that D follows from $C \wedge B$, then we can conclude that $B \Rightarrow D$

follows from C alone (and also that $C \Rightarrow D$ follows from B alone since $C \wedge B \equiv B \wedge C$); conversely, if we know that $B \Rightarrow D$ follows from C, then we can conclude that D alone follows from the composite hypothesis $C \wedge B$. In particular, since $(\text{true} \wedge B) \equiv B$,

$$\frac{\text{true} \vdash B \Rightarrow D}{B \vdash D}$$

holds for any two statements B, D; this motivates the frequent abuse of notation whereby we often write $\Rightarrow$ when we mean $\vdash$. This is an abuse that causes no major problems when no more than one implication is involved, but (for example) when we assert

$$(B \Rightarrow D) \vdash E$$

we are not asserting that E is necessarily true but only that it follows whenever $B \Rightarrow D$ is true, which in turn means .... But in particular the preceding discussion indicates the basic strategy for proving the *truth* of statements of the form $B \Rightarrow D$, namely, we *temporarily assume* that B is true and then try to show that D follows; if we succeed, then we have proven the truth of $B \Rightarrow D$ with *no* assumptions, for that is what $B \Rightarrow D$ means.

The $\Rightarrow$ operator has many properties:

Exercise A.5

$$
\begin{array}{lll}
\text{true} & \vdash & (B \Rightarrow B) & \text{for all } B \\
B \wedge (B \Rightarrow D) & \vdash & D & \text{for all } B, D \\
(A \Rightarrow B) \wedge (B \Rightarrow C) & \vdash & (A \Rightarrow C) & \text{for all } A, B, C \\
\text{if } B_1 \vdash B_2 \text{ then } (A \Rightarrow B_1) & \vdash & (A \Rightarrow B_2) & \text{for all } A \\
\text{if } A_2 \vdash A_1 \text{ then } (A_1 \Rightarrow B) & \vdash & (A_2 \Rightarrow B) & \text{for all } B
\end{array}
$$

$$
\begin{array}{lll}
((C \wedge B) \Rightarrow D) & \equiv & (C \Rightarrow (B \Rightarrow D)) & \text{for all } C, B, D \\
B \Rightarrow (D_1 \wedge D_2) & \equiv & (B \Rightarrow D_1) \wedge (B \Rightarrow D_2) & \text{for all } B, D_1, D_2 \\
(B_1 \vee B_2) \Rightarrow D & \equiv & (B_1 \Rightarrow D) \wedge (B_2 \Rightarrow D) & \text{for all } B_1, B_2, D \\
(\text{true} \Rightarrow D) & \equiv & D & \\
(\text{false} \Rightarrow D) & \equiv & \text{true} & \\
(D \Rightarrow \text{true}) & \equiv & \text{true} &
\end{array}
$$

$\lozenge$

The expressions above are all to be proved using the basic rules of inference for $\Rightarrow$. The presence of the operator $\Rightarrow$ satisfying its rule of inference enables us to prove an important "distributivity" between "and" and "or" that does not explicitly mention $\Rightarrow$ but which cannot be proved without it:

$$A \wedge (B_1 \vee B_2) \equiv (A \wedge B_1) \vee (A \wedge B_2)$$

Recall that the equivalence "$\equiv$" is a metastatement (statement about statements) whose assertion means that in fact $\vdash$ holds in both directions; using $\Rightarrow$ we can define an analog among statements themselves:

$$B \Longleftrightarrow D$$

is an abbreviation that stands for the compound statement

$$(B \Rightarrow D) \wedge (D \Rightarrow B)$$

Often $B \Longleftrightarrow D$ is read "B iff D," and the strategy for proving its truth has two parts: first temporarily assume B and deduce D; then forget that, temporarily assume D, and try to prove B (this second part is called "proving the converse" since, independently of their truth, $D \Rightarrow B$ is called the converse of $B \Rightarrow D$); only if we succeed in proving both of these do we have a right to assert the $\Longleftrightarrow$ statement, i.e. only then have we proved $B \Longleftrightarrow D$ to be true.

The reader may have noted the absence, in the exercise listing various properties of $\Rightarrow$, of any simple equivalence for statements of the form $(B \Rightarrow \text{false})$. That is because the abbreviation

$$\neg B \equiv (B \Rightarrow \text{false})$$

defines another important logical operation read "not B" roughly speaking because the statement denying B is essentially the statement that assuming B would lead to a contradiction. From the basic rules of inference we can prove the following properties of $\neg$:

Exercise A.6

$$
\begin{aligned}
(B \wedge \neg B) &\equiv \quad \text{false} \\
\text{if } (B \wedge C) \equiv \text{false then } &C \vdash \neg B \\
\neg(B_1 \vee B_2) &\equiv \quad (\neg B_1) \wedge (\neg B_2) \\
(B \Rightarrow \neg D) &\equiv \quad \neg(B \wedge D) \\
B &\vdash \quad \neg\neg B \\
\text{false} &\equiv \quad \neg \, \text{true} \\
\text{true} &\equiv \quad \neg \, \text{false}
\end{aligned}
$$

$\Diamond$

Exercise A.7

If we suppose $(B \vee \neg B) \equiv \text{true}$ for all B (this "law of the excluded middle" is not valid in all systems of statements of interest in mathematical analysis), then we further have

$$\neg\neg B \vdash B$$

$$\neg(B_1 \wedge B_2) \equiv (\neg B_1) \vee (\neg B_2) \text{ (de Morgan's law)}$$

Of course, if B is an equation $x = y$, then we usually write

$$\neg B \equiv x \neq y$$

$\Diamond$

Remark A.8: It is often useful to introduce another logical operation $B \backslash A$ called "logical subtraction" and read "B but not A" characterized by a rule of inference dual to that for implication

$$\frac{B \backslash A \vdash C}{B \vdash A \vee C}$$

Exercise A.9
Logical subtraction satisfies a whole list of properties roughly opposite to those satisfied by $\Rightarrow$. *In case* the law of excluded middle holds, we could *define* logical subtraction as $(B \backslash A) \equiv B \wedge (A \Rightarrow \text{false})$. However, in general $\backslash$ may exist when $\Rightarrow$ is impossible and in other cases $\Rightarrow$ may exist but $\backslash$ is impossible. In terms of $\backslash$ we can define a *different* negation operator $A' = (\text{true} \backslash A)$, but even when *both* operations $(\ )'$, $\neg(\ )$ are present A' may be *different* from $\neg A$. (Example: the A's, B's, etc., denote real numbers between 0, 1, and $\vdash$ denotes $\leq$ of real numbers.) $\Diamond$

Exercise A.10
If we define

$$\partial A \equiv A \wedge A'$$

(read "boundary of A") in a system where $\backslash$ is present, then from the rule of inference follows

$$\partial(A \wedge B) \equiv ((\partial A) \wedge B) \vee (A \wedge \partial B)$$

a boundary formula of topology that is important for the boundary-value problems of analysis and physics (note the similarity with the Leibniz rule for differentiation of quantities). $\Diamond$

To specify the role of the existential quantifier $\exists$ and the universal quantifier $\forall$ it is necessary to be more specific about the kinds of statements we are considering. Consider two universes X, Y of things (points, bodies, quantities, etc.) being talked about and a mapping f between them:

$$X \xrightarrow{\ f\ } Y$$

Assume that there are two systems of statements both having some or all of the logical operations previously discussed; one system refers to the things in X, the other to the things in Y. Thus,

$$A \vdash_X C$$

means that both A, C refer to the things in X and that whenever A is true of something in X, then C is also true of the same thing; this is frequently symbolized by the use of a variable x to denote the things in X, and $A(x)$ signifies that x satisfies the statement A. Then

$$A(x) \vdash_x C(x)$$

means the same as

$$A \vdash_X C$$

i.e., any x in X that satisfies A also satisfies C. Similarly,

$$B \vdash_Y D$$

means the same sort of relation between statements about the things in Y. Now the role of the given mapping f is this: it associates to every thing x in X a specific thing fx in Y. Hence, to any statement B about the things in Y there is a corresponding statement Bf about things in X

$$(Bf)(x) = B(f(x))$$

by definition. This process is often called "substitution" or "inverse image" along f. For example, if $f(x) = x^2$ and $B(y) \equiv (y - 1 = 10)$, then $(Bf)(x) \equiv (x^2 - 1 = 10)$, whereas if $D(y) \equiv (y + 1 = 0)$, then $(Df)(x) \equiv (x^2 + 1 = 0)$. It is easy to see that

$$\text{if } B \vdash_Y D \quad \text{then } Bf \vdash_X Df$$

for any statements B, D about things in Y. Thus, the process of substitution of f into statements goes "backwards" relative to f, i.e., it takes statements on the codomain Y of f back to statements on the domain X of f, and moreover it "preserves entailment $\vdash$," i.e. maps statements in fact related by $\vdash_Y$ to statements related by $\vdash_X$.

Both operations $\forall_f$ and $\exists_f$ by contrast are "covariant," meaning that they go in the same direction as f. Thus, in many situations, for any admissible statement A on X, the statement $\exists_f A$ will be an admissible statement on Y, where $\exists_f$ is related to substitution along f by the following rule of inference

$$\frac{\exists_f A \vdash_Y B}{A \vdash_X Bf}$$

for any admissible statement B on Y, or using the "variable" notation,

$$(\exists_f A)(y) \vdash B(y) \quad \text{for any } y \text{ in } Y$$
$$\text{iff}$$
$$A(x) \vdash B(f(x)) \quad \text{for any } x \text{ in } X$$

This rule of inference is precisely the rule for carrying out calculations on the basis of the meaning of $\exists_f A$, which is

$$(\exists_f A)(y) \equiv \begin{cases} \text{"there exists something } x \text{ in } X \\ \text{that satisfies } A(x) \text{ } and \\ \text{goes to } y \text{ under } f : f(x) = y\text{"} \end{cases}$$

and is clearly a statement about y's rather than x's. For (going from below the line to above) if A, B are related by $A(x) \vdash_X B(f(x))$ for all x, then to infer $B(y)$ for a given y, it suffices to know that we can find *some* x for which $y = f(x)$ and for which $A(x)$ holds; (going down) if conversely the existential statement entails B on Y for all y, then the entailment below the line must hold for all x.

The expression $\exists_f A$ names the *image* of A along f, denoted $f(A)$.

Exercise A.11

$$\text{If } A_1 \vdash_X A_2 \text{ then } \exists_f A_1 \vdash_Y \exists_f A_2$$
$$\exists_f \text{ false}_X \equiv_Y \text{ false}_Y$$
$$\exists_f (A_1 \vee A_2) \equiv_Y (\exists_f A_1) \vee (\exists_f A_2)$$
$$A \vdash_X (\exists_f A)f$$
$$\exists_f (Bf) \vdash_Y B$$

$\Diamond$

Similarly, the universal quantification $\forall_f$ along f is characterized by its relation to substitution along f, which is the rule of inference

$$\frac{B \vdash_Y \forall_f A}{Bf \vdash_X A}$$

for all statements A on X and B on Y. This rule governs the calculations appropriate to the meaning

$$(\forall_f A)(y) \equiv \text{"for all things } x \text{ in } X \text{ for which } fx = y, A(x) \text{ holds"}$$

Exercise A.12
Formulate and prove the basic properties of $\forall_f$ that are dual to those stated in Exercise A.11 for $\exists_f$.

$\Diamond$

We now give three basic examples of mappings f along which quantification is often performed. First, suppose $Y = 1$ is a universe with only one thing but X is arbitrary. Then only one mapping f

$$X \longrightarrow 1$$

is possible. In the category of abstract, constant sets (which one usually refers to as "the" category of sets), 1 has only two subsets, so that up to equivalence there are only two statements on 1, "true" and "false". Their inverse images under the unique f are true_X and false_X, but on X there are in general many more statements $A(x)$ possible (except if X is the *empty* universe, in which case $\text{true}_X = \text{false}_X$). A general A really depends on x in the sense that $A(x_0)$ may hold for one thing x_0 in X, and $A(x_1)$ may not hold for another thing x_1 in X. But $\exists_f A$ no longer depends on anything; it is either absolutely true (in case $A(x)$ holds for at least one x) or absolutely false (in case $A(x)$ fails for all x); in other words

$$\exists_f A \equiv_1 \text{true}_1, \ A \not\equiv \text{false}_X$$

$$\exists_f A \equiv_1 \text{false}_1, \ A \equiv \text{false}_X$$

For a more general and typical example, suppose Y is an arbitrary set (or "universe of things") but that we have another set T and define X to be the cartesian product $T \times Y$. Then, a simple choice of f is the *projection* map

$$
\begin{array}{ccc}
X & \xrightarrow{\ f\ } & Y \\
{\scriptstyle =}\big\downarrow & \nearrow{\scriptstyle \text{proj}} & \\
T \times Y & &
\end{array}
\qquad
\begin{array}{l}
f(x) = y \text{ if } x = \langle t, y\rangle \\
\text{i.e. } f(t,y) = y \text{ all } \langle t,y\rangle \text{ in } X
\end{array}
$$

where we recall that everything x in X is uniquely expressible in the form $x = \langle t, y \rangle$, where t is in T and y is in Y. Now if we consider any admissible statement B on Y, we have

$$(Bf)(t, y) \equiv B(y)$$

i.e. Bf is the "same" statement as B but is now considered to depend (vacuously) on the additional variable t. On the other hand, a typical statement A on X really depends on both t and y (and for that reason is often called a "relation" between T and Y), but its existential quantification along the projection f no longer depends on t:

$$\frac{(\exists_f A)(y) \vdash_Y B(y)}{A(t, y) \vdash_{T \times Y} B(y)}$$

Usually, a notation for f is chosen emphasizing what it *forgets* (namely t) rather than the rest of x (namely y) that it retains; this is because, when given y, the existence of an x that maps to it by the projection is equivalent to the existence of a t such that the pair $\langle t, y \rangle$ has the required property, since x for which $f(x) = y$

is uniquely of that form $x = \langle t, y \rangle$ for a t uniquely determined by x (of course t is *not* uniquely determined by y, except for special A). Thus, traditionally one writes

$$\exists t\, A \equiv \exists_f A$$

for f the projection on a cartesian product that leaves out t, and A a statement on that product, or (slightly confusingly) with all variables displayed

$$\exists t\, A(t, y) \equiv_Y (\exists_f A)(y)$$

where $f(t, y) = y$. The left-hand side depends on y but *not* on t; one says that the operator $\exists t$ "binds" the variable t just as a definite integral $\int_0^1 f(t)dt$ is merely a number, not a function of t. Similarly

$$\forall t\, A(t, y) \equiv_Y (\forall_f A)(y)$$

where f is the projection $f(t, y) = y$ and A is a statement on the entire cartesian product $X = T \times Y$ since to say that all x for which $fx = y$ have property A (y being given) is in this special case the same as to say that *all* t, when paired with y, satisfy the relation $A(t, y)$ owing to the nature of f.

Our first example is a special case of this second one, since if $Y = 1$ then $X = T \times Y = T$. A further important subcase of this second example is as follows: suppose $Y = T^2 = T \times T$; then $X = T \times Y = T \times T^2 = T^3$, and the same projection f is now

$$f(t_1, t_2, t_3) = \langle t_1, t_2 \rangle$$

The statements B on Y are binary relations on T, whereas the statements A on X are ternary relations on T. Quantifying along this f (either $\forall$ or $\exists$) a ternary relation gives a binary relation. As an example of how this is used in expressing properties of particular binary relations B on T (i.e., statements on T^2), let us consider a particular one we will denote by $\leq$:

$$t_1 \leq t_2 \equiv_{T \times T} B(t_1, t_2)$$

The notation suggests that we might want to consider whether B is "transitive", a condition usually written as

$$t_1 \leq t_2 \wedge t_2 \leq t_3 \Rightarrow t_1 \leq t_3$$

As mentioned before, we do not really need the complication of the $\Rightarrow$ *operator* to assert something as simple as this; the $\vdash$ relation on statements is good enough, but *which* $\vdash$? Clearly, it must be the $\vdash_3$ between statements on $X = T^3$ since the left-hand side of the implication involves three variables, but then what is the meaning of the two sides when all we were given concretely to speak about was the *binary* relation B, that is, a certain statement on $Y = T^2$? There are three different tautological f's that enter essentially into the construction even if we do not always explicitly mention them,

$$X = T^3 \underset{\longrightarrow}{\overset{\longrightarrow}{\longrightarrow}} T^2 = Y$$

namely, we define

$$\pi_3(t_1, t_2, t_3) = \langle t_1, t_2 \rangle$$
$$\pi_1(t_1, t_2, t_3) = \langle t_2, t_3 \rangle$$
$$\pi_2(t_1, t_2, t_3) = \langle t_1, t_3 \rangle$$

following the custom to name the projection for what it leaves out. Then with B denoting the given $\leq$ relation on T^2,

$$(B\pi_3)(t_1, t_2, t_3) \equiv t_1 \leq t_2$$
$$(B\pi_1)(t_1, t_2, t_3) \equiv t_2 \leq t_3$$

are both statements on $X = T^3$ so that the $\wedge$ operator for such statements can be applied by forming

$$(B\pi_3 \wedge B\pi_1)(t_1, t_2, t_3) \equiv (t_1 \leq t_2) \wedge (t_2 \leq t_3)$$

and the result compared with the other $B\pi_2$. The expression

$$B\pi_3 \wedge B\pi_1 \vdash_3 B\pi_2$$

is then the entailment relation signifying that B on T^2 is transitive. But now noting that $B\pi_2$ is actually independent of the middle variable t_2, we can (by the rule of inference for $\exists$) existentially quantify to get an entailment *on T^2*,

$$\exists \pi_2(B\pi_3 \wedge B\pi_1) \vdash_2 B$$

which *equally well* expresses the same property of B; or in variable notation,

$$\exists r[t \leq r \wedge r \leq s] \vdash_2 t \leq s$$

That is, to prove $t \leq s$ by use of transitivity is independent of any *specific r* between them, it is sufficient to know that "there exists" one.

Both of the preceding forms of the transitivity condition on B are used in everyday reasoning; it is essential to know that they are equivalent. However, one could say crudely (and a similar comment can be made about each of the logical operators): although the determining rule of inference says that the operator is "not needed" when it occurs on one side of the $\vdash$, once the operator is determined it can also occur on the other side of the $\vdash$; there it usually is essential and cannot be eliminated. For example, we could express $C \vdash D_1 \wedge D_2$ without $\wedge$ just by listing $C \vdash D_1, C \vdash D_2$, but $A \wedge B \vdash C$ expresses (except in trivial cases) that A, B must be jointly assumed to deduce C; of course in that situation we can "eliminate" the $\wedge$ in favor of the $\Rightarrow$, $A \vdash (B \Rightarrow C)$, but we can in general go further: i.e. having introduced $\Rightarrow$, we have to admit the possibility of conditions like $(B \Rightarrow C) \vdash D$, but there is no way in general to eliminate this occurrence of $\Rightarrow$. Similarly $C_1 \vee C_2 \vdash D$ is equivalent to listing $C_i \vdash D$, but a condition like $C \vdash D_1 \vee D_2$ cannot be expressed in general in terms of simpler operators. For example, the condition that a transitive binary relation $\leq$ be a *linear* order

$$\text{true}_{T^2} \vdash_{T^2} [t_1 \leq t_2 \vee t_2 \leq t_1]$$

which is just

$$\text{true} \vdash_2 B \vee B\sigma \qquad \sigma(t_1, t_2) = \langle t_2, t_1 \rangle$$

in the notation used above, cannot be expressed in terms of equations or simple logical operations such as $\wedge$ but requires the use of $\vee$. Now a similar remark involving $\exists$ concerns the statement that a transitive relation be "dense". If we consider a transitive binary relation

$$t < s \equiv D(t, s)$$

(using different symbols to distinguish from B) it is said to be *dense* if (in variables)

$$t < s \vdash \exists r[t < r \wedge r < s]$$

i.e., (roughly speaking) if more things can always be found *between* given things (relative to D). Without variables, this says

$$D \vdash_2 \exists \pi_2[D\pi_3 \wedge D\pi_1]$$

For example, this property holds in the universe of real numbers but *not* in the universe of whole numbers with the usual interpretation of D. So long as we consider only the "structure" T, D by itself, and hence only statements that can be built up logically from D, there is no way to eliminate the $\exists$ (since it occurs on the *right* hand side of $\vdash$). The example of real numbers shows, however, that if we have at our disposal additional structure, in particular if we can make statements of the type

$$F(r, t, s) \equiv r = \frac{t + s}{2}$$

we can (using equational axioms, etc., about F) prove outright that if $t < s$, then

$$t < \frac{t + s}{2} \wedge \frac{t + s}{2} < s$$

from which the $\exists r \ldots$ could be deduced since we already have something more explicit.

As a third example, we will consider the case in which $X \xrightarrow{f} Y$ is the "diagonal" mapping

$$X \xrightarrow{\Delta} X \times X$$

defined by

$$\Delta(x) = \langle x, x \rangle$$

Then, for any binary relation B on X (statement on X^2)

$$(B\Delta)(x) \equiv B(x, x)$$

defines a unary relation (statement on X). Special interest attaches to

$$E \equiv \exists_\Delta(\, \mathrm{true}_X)$$

the binary relation on X obtained by existentially quantifying the trivially true unary relation along the diagonal map. By the rule of inference for $\exists$,

$$\frac{E \vdash_2 B}{\mathrm{true}_X \vdash_1 B\Delta}$$

or in terms of variables,

$$E(x_1, x_2) \vdash B(x_1, x_2)$$

holds if and only if B is a binary relation whose diagonalization is outright true, that is,

$$\mathrm{true}_X \vdash B(x, x)$$

To obtain a class of B with that property, consider any (unary) statement A on X and consider

$$B(x_1, x_2) \overset{\text{def}}{=} (A(x_1) \Rightarrow A(x_2))$$

i.e. $B \equiv (Ap_1 \Rightarrow Ap_2)$, where $p_i(x_1, x_2) = x_i$ $\quad i = 1, 2$; then

$$(B\Delta)(x) \equiv (A(x) \Rightarrow A(x))$$

which is identically true since any statement truly implies itself. Hence, by the preceding fragment of the rule of inference for $\exists$,

$$E \vdash B$$

proving

$$E(x_1, x_2) \vdash (A(x_1) \Rightarrow A(x_2)) \qquad \text{for any } A$$

which is the usual rule of inference for *equality* E,

$$x_1 = x_2 \overset{\text{def}}{\equiv} E(x_1, x_2)$$

the rule being often called the rule of *substitution* for equality. Using equality E, we can make explicit the usual distinction between our B and D by

$$E \vdash_2 B$$

$$E \wedge D \vdash_2 \text{false}_{T^2}$$

which express that B is *reflexive* whereas D is *antireflexive* (stronger than merely being nonreflexive). Although antireflexivity is thus expressed using only $\wedge$ and false, we could introduce the somewhat more complicated operator $\Rightarrow$ if we wished by quoting *its* rule of inference, yielding

$$D \vdash (E \Rightarrow \text{false})$$

By the definition of "not," we see that (when we can use the operator $\Rightarrow$) antireflexivity of D is equivalent to

$$D \vdash \neg E$$

or with variables

$$t_1 < t_2 \vdash t_1 \neq t_2$$

or if one prefers (since we used $\Rightarrow$ anyway)

$$\text{true} \vdash t_1 < t_2 \Rightarrow t_1 \neq t_2$$

Finally, since one usually treats the "true $\vdash$" as understood when asserting something,

$$t_1 < t_2 \Rightarrow t_1 \neq t_2$$

An extremely important use of the equality relation E is in the definition of the *unique existential* quantifier that establishes a link between the general relations expressed by "statements," as discussed above on the one hand and the well-defined "operations" of algebra and analysis on the other hand. First, we consider uniqueness (without commitment about existence). If A is any statement on T, we want to express the idea that at most one thing in T satisfies A. This would mean that whenever

t_1, t_2 both have the property expressed by A, then in fact t_1, t_2 denote the same thing (the traditional example is $A(t) \equiv$ "t is the king of France at a given time"); in other words,

$$A(t_1) \wedge A(t_2) \vdash t_1 = t_2$$

where the entailment $\vdash$ is obviously supposed to be on $T \times T$. It is important to allow additional variables, so we generalize this to

$$A(t_1, y) \wedge A(t_2, y) \vdash_{T^2 \times Y} t_1 = t_2$$

where the right-hand side must mean $E\pi$, when the map $T^2 \times Y \xrightarrow{\pi} Y$ is the obvious projection, and the left-hand side is $A\pi_1 \wedge A\pi_2$ where the maps

$$T^2 \times Y \underset{\pi_2}{\overset{\pi_1}{\rightrightarrows}} T \times Y$$

are again obvious once we recognize that there are two of them. Finally, it is very important to be able to consider this uniqueness property of A independently of whether we know it is true, so we use $\Rightarrow$ instead of $\vdash$ to define a *new statement,* *"uniqueness in T of A,"*

$$\mathrm{Un}_T A \equiv \forall_\pi [A\pi_2 \wedge A\pi_1 \Rightarrow E\pi]$$

which is a statement on Y whenever A is a statement on $T \times Y$. With variables,

$$(\mathrm{Un}_T A)(y) \equiv \forall t_1, t_2 [A(t_1, y) \wedge A(t_2, y) \Rightarrow t_1 = t_2]$$

For example $A(t, y)$ might have the meaning "t is a solution of a certain differential equation with initial value y". Then, to prove $\mathrm{Un}_T A$ for y would mean that any two solutions of the equation having initial value y are equal (as functions); this is a separate issue from that of whether there are any solutions starting from y, and both are independent (unless we know more about the differential equation) from the issue of whether existence or uniqueness holds for some other y.

Exercise A.13

Let $T = \mathbb{R} = Y$, i.e. both universes are the set of all real numbers. Consider the relation

$$A(t, y) \equiv [y = t(t-1)]$$

Then

$$\vdash (\mathrm{Un}_T A)(y) \iff y \le -\frac{1}{4}$$

For example,

$$(\mathrm{Un}_T A)(0) \vdash \text{false}$$

(Draw graph.) $\Diamond$

We have already discussed

$$(\exists_T A)(y) \equiv \exists t\, A(t, y)$$

and so we can define

$$\exists!_T A \equiv \exists_T A \wedge \mathrm{Un}_T A$$

a statement on Y, for any statement A on $T \times Y$, written in variables

$$\exists! t\, A(t, y) \equiv \exists t[A(t, y)] \wedge \forall t_1, t_2[A(t_1) \wedge A(t_2) \Rightarrow t_1 = t_2]$$

and read "there exists a unique t for which $A(t, y)$".

Exercise A.14
With the notation of the previous exercise,

$$\vdash (\exists_T A)(y) \Longleftrightarrow y \geq -\frac{1}{4}$$

and

$$\vdash (\exists!_T A)(y) \Longleftrightarrow y = -\frac{1}{4}$$

(Refer to the graph of the function defined in the previous exercise.) ◇

Now if $Y \xrightarrow{g} T$ is any given mapping, we can define a statement G on $T \times Y$ (often called the "graph" of g) by

$$G(t, y) \equiv [t = g(y)]$$

Then it will be true (as a statement on 1) that

$$\forall y \exists! t\, G(t, y)$$

Conversely, (by the very meaning of "arbitrary" mapping) if we have any statement G on $T \times Y$ for which the above $\forall \exists!$ statement on 1 is true, there will be a mapping $Y \xrightarrow{g} T$ whose graph is G. What will be its value at any given thing y in Y? It will be "the" t (justified by the ! clause) for which $G(t, y)$ (which exists by the $\exists$ clause in the $\exists!$).

Exercise A.15

If G is a statement on $T \times Y$ and H is a statement on $X \times T$, then their *composition* is defined to be the statement $X \times Y$ expressed by

$$\exists t [H(x, t) \wedge G(t, y)]$$

This is actually $\exists_\pi [H\pi \wedge G\pi]$, where each of the three π's are *different* mappings with domain $X \times T \times Y$. A binary relation on T is transitive if and only if it follows from its composition with itself. If G, H are graphs of mappings $Y \longrightarrow T, T \longrightarrow X$, then their composition is the graph of a mapping $Y \longrightarrow X$. ◊

Exercise A.16

The condition that G be the graph of a mapping can be expressed *without* using $\forall, \Rightarrow$ by using instead $\exists, \wedge$. In fact, we can use $\exists, \wedge$ in the simple combination known as composition; namely, consider the transpose relation G^* of G defined by

$$G^*(y, t) \equiv G(t, y)$$

and consider E_T, E_Y the equality relations on T and Y, respectively; then

> G is the graph of a mapping only if the two entailments
> $$E_Y \vdash_{Y^2} G^* \circ G$$
> $$G \circ G^* \vdash_{T^2} E_T$$
> hold

where the small circle denotes composition. (In fact if for a given G there is *any* relation G^* that satisfies these two entailments, then G^* must be the transpose of G, as defined above, and hence G is the graph of a mapping.) This reformulation is important since $\exists, \wedge$ (which go into the definition of $\circ$) are stable under many more geometrical transformations used in analysis than are $\forall, \Rightarrow$ (which go into the definition of !). Moreover this latter kind of relation (expressed in the box) between "relations" and "mappings" will persist in situations in which the values of $G(t, y)$ are much richer mathematical objects than just "yes" and "no" answers. ◊

A.2 Fields, Nilpotents, Idempotents

Examples of Logical Operators in Algebra

The proofs of many of the exercises in this section will be clearer if explicit rules of inference from the previous section are cited at the appropriate points.

The most basic properties of algebraic structures such as rings, linear spaces, and categories are expressed by equations; for example, the distributive property, nilpotence, associativity, or the property of being a solution. However, in working with these equations we must frequently use stronger logical operators, both in stating stronger axioms on the ground ring in linear algebra and in summarizing the meaning of our complicated calculations. (It should be remarked, however, that most of this logic again becomes equational when we pass to a higher realm.) For example, using the logical symbol $\vdash$, which can be read "entails," the additional axioms stating that a given ring R is a **field** are that R is nondegenerate,

$$0 = 1 \vdash \text{false}$$

(usually expressed by introducing "not" and saying

$$\text{true} \vdash 0 \neq 1)$$

and that every nonzero quantity in R has a reciprocal,

$$x \neq 0 \vdash \exists y[xy = 1]$$

When the law of excluded middle is valid, the latter is equivalent to the (in general stronger) condition, involving the logical symbol for "or,"

$$x = 0 \vee \exists y[xy = 1]$$

being true$_R$ (which has the virtue of being invariant under more geometrical transformations but the drawback, in those cases like continuous functions where the law of excluded middle is false, of being less likely to be true). Usually, one expresses this field axiom using $\forall \Rightarrow$ as

$$\text{true} \vdash \forall x[x \neq 0 \Rightarrow \exists y[xy = 1]]$$

with the understanding that the universe over which both x, y vary is R. Thus $\mathbb{Z} = \{\ldots -3, -2, -1, 0, 1, 2, 3 \ldots\}$ is a ring R that is not a field since, for example, $5 \neq 0$, but there is no y in $\mathbb{Z}$ for which $5y = 1$. In any ring we can deduce purely equationally from the hypotheses

$$xy_1 = 1$$
$$xy_2 = 1$$

that $y_1 = y_2$ (here is the deduction, using only [commutative] ring axioms and the hypotheses:

$$y_1 = y_1 1 = y_1(xy_2) = (y_1 x)y_2 = (xy_1)y_2 = 1 y_2 = y_2)$$

Therefore, we can conclude that in any ring, reciprocals are unique, and hence in any field that (using the exclamation point to signify this uniqueness)

$$\forall x[x \neq 0 \Rightarrow \exists! y[xy = 1]]$$

Further (since R is nondegenerate if it is a field), "the" (just justified) reciprocal of x cannot be zero either.

Exercise A.17

If y is a reciprocal of x, then x is a reciprocal of y; if, in any given ring R, 0 has a reciprocal, then R is degenerate. In case R is a field, if we restrict the universe to the set G of all nonzero elements of R (G is no longer a ring) the slightly simpler statement

$$\forall x \exists! y[xy = 1]$$

is true over G. Since this is the criterion for the existence of a mapping, there is a mapping

$$G \xrightarrow{\;()^{-1}\;} G$$

called the reciprocal mapping whose graph is defined by

$$x \cdot y = 1$$

that is, $y = x^{-1}$ if and only if $x \cdot y = 1$.   ◊

Exercise A.18

A much better way of understanding the last construction is as follows: Let R be any commutative ring (not necessarily a field, maybe even degenerate). Define G to be the subset of R consisting of all elements x of R satisfying

$$\exists y[xy = 1]$$

in R. Then there is a reciprocal mapping $G \longrightarrow G$, 1 is in G, and G is closed under multiplication; i.e., if x_1, x_2 are in G, then $x_1 x_2$ is in G since $1^{-1} = 1$, and $(x_1 x_2)^{-1} = x_2^{-1} x_1^{-1}$. This means that G is a (commutative) group called "the multiplicative group of R". If 0 is in G, then R is degenerate. For any x in G, $-x$ is also in G. But x_1, x_2 in G does not imply $x_1 + x_2$ in G. If $R = \mathbb{R}$, the real numbers, then $1 + x^2$ is always in G for any x, and the same is true if $R = C(S) =$ the ring of all continuous real-valued functions on any continuous domain ("topological space") S.   ◊

Now the condition that a ring R be a field is just that R be the disjoint union of $\{0\}$ and G, namely that (reading the $\vee$ form of the definition backwards and using the symbol for "not")

$$\exists! x[\neg G(x)]$$

Since any ring R has a special element 1 and since R has an addition operation, there are elements in R that may as well be denoted

$$2 = 1 + 1$$
$$3 = 1 + 1 + 1$$
$$\vdots$$

(not all of these need be distinct). Even if R is a field, not all of these need have reciprocals; for example, there is an important field with only three elements in all in which $3 = 0$. However, a great many rings, even those that are not fields, do involve $I\!R$ in such a way that all of the above do have reciprocals, which can be denoted as usual by $\frac{1}{2}, \frac{1}{3}, \ldots$ Thus, $\frac{1}{2}$ is in G, $\frac{1}{3}$ is in $G \ldots$, where G denotes the multiplicative group of any such ring R.

Exercise A.19

In any ring having $\frac{1}{2}$,

$$\text{true} \vdash \forall x \exists y [y + y = x]$$

$\Diamond$

By a **subring** of a given ring R is meant a subset S of the elements of R that contains 0, 1 and is closed under the addition, the subtraction, and the multiplication of R. Thus, if p is any polynomial with coefficients in $\mathbb{Z}$ in several, say three, variables, and if x, y, z are in S, then $p(x, y, z)$ is also in S.

Exercise A.20

If R is a ring having $\frac{1}{2}$ and if S is a subset containing 0, 1 closed under addition and closed under the unary operations of multiplication by $\frac{1}{2}$ and by -1, then S is a subring if and only if S is closed under the unary operation of squaring. (The answer is a frequently used formula.) $\Diamond$

Now a subring is not necessarily closed under division, even to the extent to which division is defined in R. Thus, $\mathbb{Z} \subset I\!R$ is a subring, but $\mathbb{Z}$ is not a field even though $I\!R$ is a field. But any subring of any field does have a special property not shared by all rings of interest, namely

$$xy = 0 \vdash [x = 0 \lor y = 0]$$

Exercise A.21

Prove the statement just made in any subring of a field. $\Diamond$

A nondegenerate ring S having this property for all x, y in S is called an **integral domain**. This is intimately related to the cancellation property

$$\forall x_1, x_2[ax_1 = ax_2 \Rightarrow x_1 = x_2]$$

for an element a, which (using subtraction) is easily proved equivalent to the "non-zerodivisor" property of a

$$\forall x[ax = 0 \Rightarrow x = 0]$$

where all universal quantifiers range over all x, x_1, x_2 in the ring in which we are considering a. We might call a "monomorphic" in that ring. Note that this property uses the logical operators in an essential way since, when we want to prove

$$a \text{ is monomorphic } \vdash \text{ something else about } a$$

we cannot always eliminate the $\forall$, $\Rightarrow$ implicit on the left-hand side. Of course, if the "something else" is just another instance of the cancellation property, such proof may present no problem. Now clearly $a = 0$ can not be monomorphic in a nondegenerate ring since

$$ax = 0 \cdot x = 0$$

for any x, yet we could take $x = 1$; hence, it would not follow that $x = 0$. Now the idea of an integral domain is that (if the law of excluded middle is assumed) the only a that is not monomorphic in R is $a = 0$. That is, the validity for all a in R of any one of

$$a \neq 0 \Rightarrow \forall x[ax = 0 \Rightarrow x = 0]$$
$$\exists x[ax = 0 \wedge x \neq 0] \Rightarrow a = 0$$
$$\forall x[ax = 0 \Rightarrow [a = 0 \vee x = 0]]$$

is equivalent (using the law of the excluded middle) to the condition that R is an integral domain. The last form with $\vee$ is the one familiar from high school as a crucial step in the method of solving polynomial equations by factoring. This method is used in proving

Theorem A.22: *In any integral domain, the equation*

$$x^2 = x$$

has precisely two solutions.

Proof: If $x^2 = x$ then $x^2 - x = 0$, and hence $x(x - 1) = 0$ (since $x(x - 1) = x^2 - x$ in any ring). Now use the integral domain property to get $x = 0 \vee x - 1 = 0$, i.e. $x = 0 \vee x = 1$. We say "precisely two" because the ring is nondegenerate. ∎

Exercise A.23

In any commutative ring, an element satisfying $x^2 = x$ is called **idempotent**. If x is an idempotent, then so is its "complement" $1 - x$. The product of any two idempotents is an idempotent. If x and y are idempotents and if $xy = 0$ (one says x, y are "disjoint" or "orthogonal") then $x + y$ is also an idempotent. One says a ring "has connected spectrum" if it has precisely two idempotents. In general, the idempotents describe chunks of the "spectrum," for example, of a linear transformation (which gives rise to a ring). ◊

Very important in engineering calculus [B99], in analyzing linear transformations, and so on, are the nilpotent elements in commutative rings, where x is **nilpotent** if and only if

$$\exists n [x^{n+1} = 0]$$

Here the $\exists n$ does not range over the ring we are talking about but rather over the set $0, 1, 2, 3 \ldots$ of natural numbers, which *act as exponents* on elements of any ring (or indeed of any system wherein at least multiplication is defined). In more detail, we could say that x is nilpotent of order 1 if

$$x^2 = 0$$

while $x \neq 0$, that x is nilpotent of order 2 if

$$x^3 = 0$$

while $x^2 \neq 0$, and so on. Of course 0 is nilpotent of order zero. In a nondegenerate ring $x = 1$ is not nilpotent of any order. Using commutativity, we find that the product of a nilpotent with any element is again nilpotent. Again using commutativity, we discover that the sum of any two nilpotent elements is nilpotent; however, the order may increase. For example, if $x^2 = 0$ and $y^2 = 0$, we can calculate that $(x + y)^3 = 0$; as for $(x + y)^2$, it might be 0, but only in case $xy = 0$, which is not always true. Analysis of the calculation leads to the idea that, to be sure of the nilpotency of a sum, we have to add the orders of nilpotency of the summands:

Theorem A.24: *If* $x^{n+1} = 0$, $y^{m+1} = 0$ *in a commutative ring, then always*

$$(x + y)^{n+m+1} = 0$$

Proof: In any commutative ring the distributive law implies the binomial expansion

$$(x + y)^p = \sum_{i+j=p} C_{ij} x^i y^j$$

for any x, y in the ring and any p in $\mathbb{N}$ (note that $\mathbb{N} \subset \mathbb{Z}$ and that $\mathbb{Z}$ can be used as coefficients in any ring – indeed in any system in which addition and subtraction are defined. In fact

$$C_{ij} = \frac{(i+j)!}{i!j!}$$

is in $\mathbb{N}$ despite the denominators, where ! denotes "factorial," by Pascal). Thus, the proof of the theorem reduces to the following fact about the elementary arithmetic of $\mathbb{N}$: ∎

Lemma A.25: $i + j = n + m + 1 \ \vdash \ [i \geq n + 1] \vee [j \geq m + 1].$

Exercise A.26
Prove Lemma A.25. ◇

In any case, since

$$x^{n+1} = x^n \cdot x$$

is a product, it is immediately clear that

Theorem A.27: *In an integral domain, the only nilpotent element is 0.*

An extremely important property (for analysis, linear algebra, computer science, etc.) is the following, showing that, although the existence of nilpotent elements has the "negative" consequences that some elements (the ones "near" zero) are definitely not invertible, it also has the "positive" consequence that some other elements (those "near" 1) definitely are invertible, and there is even a specific formula for the reciprocals.

Theorem A.28: *If h is any nilpotent element in a ring, then $1 - h$ has a reciprocal in the same ring. In fact if $h^{n+1} = 0$, then*

$$\frac{1}{1-h} = \sum_{k=0}^{n} h^k$$

Proof: Calculate that the right-hand side, multiplied by $1 - h$, gives 1. ∎

Remark A.29: In a ring furnished with a notion of convergence, Theorem A.28 can often be generalized to some h for which h^n merely converges to 0 as n tends to ∞, i.e. to small h's not necessarily so small as to be "nilpotent". But the formula of the theorem is surprisingly often useful even just for the nilpotent case.

Exercise A.30

If u has a reciprocal and h is nilpotent then $u \pm h$ has a reciprocal. (A formula, only slightly more complicated than that of the theorem, can either be deduced from the theorem as a corollary or calculated and proved directly.) ◊

Exercise A.31

If $u_1 = 1 - h_1, u_2 = 1 - h_2$ are invertible elements of a (commutative) ring of the form indicated with h_1, h_2 nilpotent (with orders of nilpotency n_1, n_2 say), the product $u_1 u_2$ is of course invertible; is it again of the special form, namely "infinitesimally near 1" in the sense that

$$u_1 u_2 = 1 - h$$

for some nilpotent h of some order? Start with the special case $h_1^2 = 0 = h_2^2$, $h_1 h_2 = 0$. What if $h_i = t_i \epsilon$, where $\epsilon^2 = 0$? What if multiplication were non-commutative? ◊

Remark A.32: (An Embedding) Any given integral domain R can be realized as a subring of a field F by constructing F to consist of equivalence classes of fractions $\frac{x}{s}$, where x is in R, s is in R, and $s \neq 0$.

The condition that a ring R has "no" (i.e., no nonzero) nilpotent elements is often referred to in geometry and analysis by saying the R is "reduced". It is more general than the cancellation (i.e., integral domain) property since, for example, $R = \mathbb{R}^2$ with coordinatewise multiplication is reduced (i.e., has no nilpotent element) but is not an integral domain since it has nontrivial idempotent elements $\langle 0, 1 \rangle$, $\langle 1, 0 \rangle$. In logical notation with variables, R is reduced if and only if

$$\exists n[x^{n+1} = 0] \vdash x = 0$$

holds for all x in R. Since the $\exists$ occurs on the left, this is one of its eliminable cases. But more profoundly (i.e., using something of the quantitative content of the theory of rings and not merely logical form):

Exercise A.33

If a ring satisfies

$$\vdash \forall x[x^2 = 0 \Rightarrow x = 0]$$

then it is reduced.

Hint: Show that if $x^{n+1} = 0$, then $x^{2n} = 0$; hence, using our main assumption, then also $x^n = 0$. By induction, the n can be decreased until eventually $n = 1$. ◊

Appendix B

The Axiom of Choice and Maximal Principles

The axiom of choice was first formulated by Zermelo in 1904 and used to prove his Well-Ordering theorem. The axiom was considered controversial because it introduced a highly nonconstructive aspect that differed from other axioms of set theory. For some time it was mainly used in the form of the Well-Ordering theorem (which is actually equivalent to the choice axiom). In this formulation, the axiom permits arguments by the so-called transfinite induction. For about the last 50 years it has usually been used in the form of the Maximal Principle of Zorn (published in the 1930s, though a version by Hausdorff was published earlier.)

If nontrivial variation with respect to some category S of abstract sets and arbitrary mappings is allowed in a category of variable sets, the axiom of choice tends not to hold as we have seen in Exercises 4.54, 6.12, and Section 10.3. In fact, the axiom is valid in certain very special toposes of variable sets determined relative to a category of abstract sets and arbitrary mappings, as was mentioned in Section 4.6.

Here we will demonstrate that Zorn's Maximal Principle is equivalent to the axiom of choice. We will show this for forms of the Maximal Principle that use both *chains* and *directed* (or filtered) posets. The latter form is more suitable for arguments made in mathematical practice, whereas it is the former that we will see as a direct consequence of the axiom of choice. In addition we will consider Hausdorff's Maximal Principle and some other consequences of the choice axiom. Our proofs of the maximal principles will follow directly from the famous Fixed-Point theorem of Bourbaki (B.15 below).

To state the results we need a few definitions. First recall the definition of *poset*, Definition 10.6, that is a category with at most one arrow between any two objects. Any set of objects of a poset E determines a full subcategory, which is also a poset, called a **subposet** of E.

Definition B.1: *Let S be a set of objects of a poset E. An object u in E is an* **upper bound** *for S if for any s in S, s $\leq$ u.*

Definition B.2: *Let S be a set of objects of a poset E. The object u is a* **least upper bound** (*abbreviated* **lub**) *or* **supremum** (*abbreviated* **sup**) *for S if it is an upper bound for S, and moreover for any upper bound u' for S we have u $\leq$ u'.*

Dually we have definitions of **lower bound** and **greatest lower bound** (abbreviated **glb**) or **infimum** (abbreviated **inf**) for a set *S* of objects of *E*.

Definition B.3: *Let D be a set of objects of a poset E. We say D is* **directed** *if it is nonempty and every two-element part of D has an upper bound in D, or equivalently*

$$\forall x, y \in D \; \exists u \in D \; x \leq u \; \& \; y \leq u$$

Exercise B.4

Show that the power set $\mathcal{P}X$ of any set X is a poset with arrows given by the inclusion relation $\subseteq$. Note that $\mathcal{P}X$ is directed as a part of itself. Indeed, *any* part of $\mathcal{P}X$ has a sup. $\quad\Diamond$

The next concept is important in the sequel.

Definition B.5: *An object m of a poset E is* **maximal** *if*

$$\forall x \; m \leq x \Longrightarrow m \cong x$$

A **minimal** object is defined dually.

An important special case is

Definition B.6: *An object m of a poset E is a* **maximum** *if m is an upper bound for all of E. A* **minimum** *object is defined dually.*

Exercise B.7

Show that a maximum is maximal. Show that a maximum object of a subposet S of E is a sup of S. $\quad\Diamond$

We can now state (the directed version) of

The Maximal Principle of Zorn

If every directed part of a poset E has an upper bound, then E has a maximal object.

We can immediately prove that Zorn's Maximal Principle implies the axiom of choice. This proof is a very typical example of the use of this principle in modern mathematics.

Theorem B.8: *The Maximal Principle of Zorn implies the axiom of choice.*

Proof: We need to begin with an epimorphism $e : A \longrightarrow B$ in $\mathcal{S}$ and find a section for it. This section should be a maximal object in some suitable poset, for this is what Zorn's principle provides. The objects of our poset E will be pairs $\langle i, s \rangle$, where $i : B_i \longrightarrow B$ is a part of B and $s : B_i \longrightarrow A$ satisfies $\forall b \in i$, $es(b) = b$. Note that it is immediate that such s must be mono. We call the pairs $\langle i, s \rangle$ *partial sections* of e. Partial sections of e form a poset E when $\leq$ is defined by

$$\langle i, s \rangle \leq \langle i', s' \rangle \text{ iff } i \subseteq i' \text{ \& the restriction of } s' \text{ to } B_i \text{ is } s$$

We will obtain a maximal object of E by showing that E has least upper bounds for directed parts and applying the Maximal Principle. It will turn out that such a maximal object provides a section. So let D be a directed part of E. Define a part $i_D : B_D \longrightarrow B$ of B by

$$\forall b \quad b \in i_D \text{ iff } \exists \langle i, s \rangle \text{ in } D \text{ \& } b \in i$$

Define $s_D : B_D \longrightarrow A$ by

$$s_D(b) = s(b) \text{ whenever } b \in i \text{ and } \langle i, s \rangle \text{ in } D$$

It is necessary to verify that s_D is well-defined by showing that if $b \in i$ and $b \in i'$, then $s(b) = s'(b)$. This is where we use that D is directed: there is an upper bound $\langle i'', s'' \rangle$ in D for $\langle i, s \rangle$, $\langle i', s' \rangle$. Thus, $i \subseteq i''$ and $i' \subseteq i''$, and $s(b) = s''(b) = s'(b)$. It is also obvious that $\langle i_D, s_D \rangle$ is an upper bound for D since

$$\forall \langle i, s \rangle \text{ in } D \quad b \in i \Longrightarrow b \in i_D$$

and so $i \subseteq i_D$.

Thus, E has a maximal object $\langle i_m, s_m \rangle$. We claim that s_m is the required section of e, that is, $B_m \cong B$. If not, there is $b_0 : 1 \longrightarrow B$ such that b_0 is not in i_m, and so

$$i_1 = \begin{cases} i_m \\ b_0 \end{cases} : B_m + 1 \longrightarrow B$$

is a part of B. Since e is epi there is $a : 1 \longrightarrow A$ such that $e(a) = b_0$. Let

$$s_1 = \begin{cases} s_m \\ b_0 \end{cases} : B_m + 1 \longrightarrow A$$

and then $\langle i_1, s_1 \rangle$ is a partial section of s. However, $i_m \subseteq i_m + 1$, but $\begin{cases} i_m \\ b_0 \end{cases}$ is not included in i_m (since b_0 is not in i_m), and this contradicts the maximality of i_m. Thus, s_m has domain B. ∎

There are many other uses of the Maximal Principle of Zorn. For example, it implies that any proper ideal in a commutative ring is contained in a maximal ideal (see Exercise B.34) and that any vector space has a basis.

Definition B.9: *A poset E is called a* **total order** (*or* **chain**) *if for objects x, y in E either $x \leq y$ or $y \leq x$. A subposet C of E, which is itself a chain is called a* **chain in E**.

Exercise B.10
Show that a totally ordered poset E is directed. ◇

With the definition we can state

The Maximal Principle of Zorn (chain version)

If E is a poset such that every chain in E has an upper bound, then E has a maximal object.

Exercise B.11
Show that the Maximal Principle of Zorn (chain version) implies the Maximal Principle of Zorn. ◇

Notice that a proof very similar to that of Theorem B.8 will show directly that the axiom of choice is implied by the chain version of the Maximal Principle.

Definition B.12: *For any set X, $\forall_X$ denotes the characteristic function $\Omega^X \longrightarrow \Omega$ of the one-element part of Ω^X whose element is the composite $X \longrightarrow 1 \xrightarrow{\text{true}} \Omega$ called* true_X *for short. Thus*

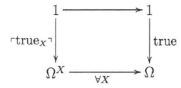

is a pullback.

Given any $I \xrightarrow{F} \Omega^X$, corresponding to an I-parameterized family of parts of X, consider its transpose τF; then, the composite $\forall_I[\tau F]$ is the characteristic function of the part of X called the *intersection* of the family F.

We will use the **higher dual distributive law** of logic

$$\forall a(La \text{ or } Ra) \text{ entails } \exists a(La) \text{ or } \forall a(Ra)$$

We will discuss below in Exercise B.26 the question of when this law is valid (it is certainly valid for any category S of abstract sets or any topos in which the axiom of choice is true).

For simplicity we will assume that posets are **strict** in the sense that $x = y$ follows from $x \leq y$ and $y \leq x$; then, sups are unique if they exist.

Definition B.13: *If f is an endomap of a set X, then a* **fixed point** *for f is an element x of X such that $fx = x$.*

Definition B.14: *If f is an endomap of the set of objects of a poset E, then an f-***chain** *is any chain C in E such that $x \in C$ implies $fx \in C$. We denote the part of the power set of the objects of E consisting of all f-chains by $ch_f(E)$.*

Theorem B.15: (Bourbaki Fixed Point theorem 1950). *Let E be a poset, and let f be an inflationary endomap of the set of objects of E, that is, for all x in E, $x \leq f(x)$. Then f has at least one fixed point provided that any chain in E has a sup in E.*

Proof: Define a *subsystem* to be any part (of the objects of E) closed with respect to sups of f-chains and to f itself. Let A be the intersection of all subsystems of E. Thus

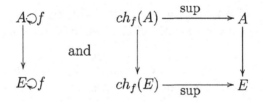

are commutative diagrams and A is the smallest part with those properties. Since we have allowed the part 0 in $ch_f(E)$, it follows that A is nonempty and indeed has a smallest element 0. Because of the first diagram, any fixed point of A is also a fixed point of E. Thus, for the remainder of the proof we work in A.

We need to establish that any subsystem (such as A) whose only subsystem is itself has the following two

Elementary Properties of Irreducible Subsystems

(P) $t \leq a$ entails $t = a$ or $f(t) \leq a$

(B)   it is true that $t \leq a$ or $f(a) \leq t$

These properties will suffice because they clearly imply the "f-trichotomy" property for A,

$$t = a \quad \text{or} \quad f(a) \leq t \quad \text{or} \quad f(t) \leq a$$

which in turn easily implies that the whole of A is an f-chain. (As often in inductive proofs, to reach the latter goal we needed to prove something stronger.) But A being an f-chain entails that $\sup A \in A$ since we have assumed that A is closed under such sups. Then $\sup A$ is a fixed point of f since $\sup A \leq f(\sup A)$, yet $x \leq \sup A$ for all $x \in A$ and $f(\sup A) \in A$.

In order to establish the two properties above, define a part of A:

$$P = \{a \in A \mid \forall t \in A[t \leq a \Rightarrow [t = a \text{ or } f(t) \leq a]]\}$$

Then, more precisely, what we will show is that $P = A$.

It will suffice to show that P is a subsystem, for then $A \subseteq P$ by definition of A. To show that P is a subsystem, we first will need to show that for any given $a \in P$ each of the following is a subsystem:

$$B_a = \{b \in A \mid b \leq a \text{ or } f(a) \leq b\}$$

Exercise B.16

Because $a \in P$, B_a is closed under f.   ◊

Exercise B.17

The subset B_a is closed under any sups that exist in A.
Thus $B_a = A$, i.e. any $a \in P$ is a "bridge point" for A.

Hint: Use the higher dual distributive law.   ◊

Both of the clauses needed to show that P is a subsystem will depend on the bridge-point property.

Exercise B.18

Assume $a \in P$. Use the fact that a is a bridge point to conclude that $fa \in P$.   ◊

Exercise B.19

Applying a different case of the higher dual distributive law to the fact that each element of P is a bridge point, conclude that P is closed under sups of f-chains.

Hint: Suppose C is any f-chain in P and t is any element of A for which $t \leq \sup C$; then it must be shown that $t = \sup C$ or $f(t) \leq \sup C$. But all elements of C are bridge points; hence, t satisfies

$$\forall a \in C[t \leq a \text{ or } f(a) \leq t]$$

The higher dual distributive law can be applied by taking $L_t(a)$ to be "$t \le a$" and $R_t(a)$ to be "$f(a) \le t$," leading to two possibilities:

(1) If $\forall a \in C[f(a) \le t]$, then $\forall a \in C[a \le t]$, and hence $\sup C \le t$, i.e. $t = \sup C$.
(2) If $t \le a$ for some $a \in C$, then since $a \in P$, one has $t = a$ or $f(a) \le t$:
  (a) if $t = a$, then $f(t) = f(a)$, yet $f(a) \in C$ since C is an f-chain, and so $f(t) \le \sup C$;
  (b) if $f(a) \le t$, together with $t \le a \le f(a)$, we have $t = f(a) \in C$, and so $f(t) \le \sup C$.

Thus, in all these cases $t = \sup C$ or $f(t) \le \sup C$; since this is true for all $t \le \sup C$, it follows that $\sup C \in P$.   ◊

Since the foregoing holds for all f-chains C, we have completed the proof that P is a subsystem. Therefore, $P = A$, and so, as shown above under "Elementary Properties of Irreducible Systems," f has a fixed point. Thus, the exercise completes the proof of the Bourbaki Fixed-Point theorem.   ■

Exercise B.20

(Hausdorff 1914) The axiom of choice implies that any poset X contains a maximal chain (the Hausdorff Maximal Principle). Indeed, if not, let φ be a section for the first projection in

$$[\text{Proper inclusions of pairs of chains in } X] \rightrightarrows \text{chains } (X)$$

and let f be φ followed by the second projection. The sup of a chain of chains always exists and is again a chain.

Note: This proof of the Hausdorff Maximal Principle evidently uses Boolean logic (which is actually a consequence of the axiom of choice, see [D75] or [Joh77]).

◊

Exercise B.21

(Zorn 1935) Any poset in which every chain has an upper bound has itself a maximal element.

Hint: An upper bound of a maximal chain is a maximal element; indeed, if C is a chain, and if m is any upper bound for C and x is any element with $m \le x$, then $C \cup \{x\}$ is a chain; thus, if C is maximal, $x \in C$ and hence $x = m$.   ◊

Combining this exercise with Theorem B.8 and Exercise B.11, we have

Theorem B.22: *The axiom of choice is equivalent to the Maximal Principle of Zorn (either version!).*

As mentioned at the beginning of this appendix, Zermelo used the axiom of choice to prove his Well-Ordering theorem.

Definition B.23: *A poset E is called* **well-ordered** *if whenever S is a nonempty part of the objects of E then S has a minimum object.*

Exercise B.24
Show that a well-ordered poset E is a total order.

Hint: Consider the minimum object of a two-object set. ◊

Exercise B.25
(Zermelo 1908) Every set X can be well-ordered.

Hint: Consider the set of all well-orderings of parts of X and order it by "initial segment"; in this case a maximal well-ordering must have the whole X as set of objects. ◊

A given surjective map has a section provided that certain associated sets are well-orderable therefore the Well-Ordering theorem is actually equivalent to the axiom of choice. In fact, there are literally dozens of useful mathematical principles in topology and algebra that have also been shown equivalent to the choice axiom over the past 75 years.

The higher dual distributive law of logic used in the above proof of the Theorem B.15 is true in many toposes but not in most. This law is dual to the usual infinite distributive law

$$u \le u_a \text{ for all } a \text{ entails } u \sum w_a \le \sum u_a w_a$$

which is true in any topos because of the existence of implication as an operation on subobjects. In a Boolean topos like $\mathcal{S}$, the power sets are self-dual, and hence the higher dual law holds as well.

Exercise B.26
Does the higher dual distributive law imply the Boolean property? Does it hold in the topos $\mathcal{S}^{\mathcal{C}^{\mathrm{op}}}$ of all actions of $\mathcal{C}$ on abstract sets, in case $\mathcal{C}$ is discrete, or a group, or a poset? ◊

Here are some additional exercises:

Exercise B.27
An irreducible f-system is not only closed with respect to sups of f-chains, but in fact has sups of all chains.

Hint: Use a third instance of the higher dual distributive law.

Exercise B.28

Any irreducible f-system, such as $P = A$ in Exercise B.19 has two further properties: The endomap f is order-preserving, and every part has a sup.   ◊

Note: The latter two properties are of importance even when f is not inflationary and the system is not irreducible: Tarski's fixed-point theorem states that every order-preserving endomap of a poset having all sups has a fixed point. The reader should be able to prove it; it is easier than the Bourbaki theorem whose consequences we are discussing in this appendix.

Exercise B.29

Use Tarski's fixed-point theorem to show that if there are parts $B \xrightarrow{\alpha} A$ and $A \xrightarrow{\beta} B$, then there exists an isomorphism $A \cong B$. This result depends on the further assumption that all parts have Boolean complements; it is not true for most categories of variable or cohesive sets.

Hint: Construct an order-preserving endomap of 2^A by using complementation in both A and B and also applying both α and β; on a fixed "point" of that endomap, use β, and on the complement use a retraction for α. The construction actually can be applied to any pair of maps; what conclusion can then be drawn?   ◊

Exercise B.30

Using Zorn's Maximal Principle (and Boolean logic) show that any family $E \longrightarrow I$ of sets makes I a chain with respect to the ordering: $i \leq j$ iff there exists an injective $E_i \longrightarrow E_j$.

Hint: If there is no injective $E_1 \longrightarrow E_2$, then any *maximal partial injection* from part of E_2 into E_1

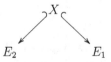

will actually have its domain total, that is $X = E_2$.   ◊

General Remark: Inductive arguments rely on special cases of the principle that within any given set E as universe, if too many conflicting higher-order operations are considered, they will be constrained to have some relation. The simplest sort of such relation is a fixed point. For example, arbitrary sups and an order-preserving endomap are sufficiently conflicting, according to the Tarski theorem, as are sups of chains and an inflationary endomap, according to the Bourbaki theorem. Some systems of operations, even higher-order ones, are sufficiently harmonious to permit "freedom" from such unexpected relations, as the following example shows:

Exercise B.31
Free sups exist. More precisely, given any set X, there is a poset PX that has sups of all parts and a map $X \xrightarrow{\eta} PX$ such that, given any map $X \xrightarrow{\varphi} E$, where E is any poset having sups of all parts (no relations are assumed in E beyond the definition of sups), there is a unique map $PX \xrightarrow{\varphi'} E$ that preserves all sups and satisfies $\varphi'\eta = \varphi$.　　　　　　　　　　　　　　◊

Of course systems of finitary operations that are free of any unexpected relations do tend to exist because of the assumption that a natural number system exists on some set. There are many variants of that theme, for example:

Exercise B.32
Consider systems E involving a nullary operation $1 \xrightarrow{\infty} E$ and a unary operation $E \xrightarrow{f} E$ subject to the condition that $f(\infty) = \infty$; a category is obtained by considering that a homomorphism between any two such systems is a map of the carrying sets that preserves both these operations. Show that there is a free such system N' generated by the one-point set, i.e. there is $1 \xrightarrow{0} N'$ such that for any such system E and for any given point $1 \xrightarrow{x} E$ there is a unique homomorphism $N' \xrightarrow{x'} E$ for which $x'(0) = x$. Does N' have any f-fixed points other than ∞? (In the category of abstract sets, N' has an ordering for which all parts have sups, but in categories of variable or cohesive sets this N' fails to have all sups.)　　　　◊

Thus, it is the combining of higher-order operations with very distinct (even finitary) ones that may be impossible to do in a "free" way within any fixed set.

Example: (Gaifman [G64]) Any infinite poset with arbitrary sups and an endomap f of the underlying set for which $ff = $ identity must satisfy some additional relation. (Actually, Gaifman showed that there are no "free" [in a sense analogous to those of Exercise B.31 and B.32] complete Boolean algebras except finite ones, which implies our statement.)　　　　　　　　　　　　　◊

The irreducible subsystems (such as A in our proof of the Bourbaki theorem) that help to reveal these relations seem to have the property that every element is somehow "reached" from below by applying the operations. It has been found useful (by Cantor and since) to make this reaching idea appear more precise by introducing well-orderings along which the operations can be "infinitely iterated". In the case of our irreducible systems $A = P$ with sups of chains and an inflationary f, this is already achieved:

Exercise B.33

Consider a chain C having sups of any part as well as an order-preserving endomap f such that $x \leq fx$ for all x. (Then there is a smallest element 0 and a largest element ∞, and moreover $f(\infty) = \infty$.) If C has no parts closed under f and sups (other than C itself), then

$$f(x) = x \Rightarrow x = \infty$$

Can you show that C is well-ordered?   ◊

Returning to finitary algebra, the following is a typical application of Zorn's Maximal Principle:

Exercise B.34

If A is any commutative ring in which it is false that $0 = 1$, then there exists a field F and a surjective ring homomorphism $A \longrightarrow F$ (such a homomorphism is a "closed point" of the space $\mathrm{spec}\,A$, which plays a key role in algebraic geometry and functional analysis. Since typically A has a great many surjective homomorphisms to other rings $A \longrightarrow B$, where also $0 \neq 1$ in B [such homomorphisms correspond to nonempty closed subspaces of $\mathrm{spec}\,A$], by applying the exercise to B and composing, we see that $\mathrm{spec}\,A$ typically has many closed points).

Hint: Surjective homomorphisms with domain A are determined by their kernels, which are A-linear subgroups of A; the codomain of such a homomorphism is a field iff the corresponding A-linear subgroup is maximal among those that do not contain 1. The A-linear subgroups have been known since Dedekind as **ideals**. ◊

Appendix C

Definitions, Symbols, and the Greek Alphabet

C.1 Definitions of Some Mathematical and Logical Concepts

Included here are some of the main definitions from the text. Several entries go beyond bare definition, in an attempt to provide a window into the historical and foundational background.

***Adjoint Functors*:**

Let $\mathcal{X}$ and $\mathcal{A}$ be categories and $F : \mathcal{X} \longrightarrow \mathcal{A}$ and $G : \mathcal{A} \longrightarrow \mathcal{X}$ be functors. We say that F **is left adjoint to** G (or equivalently G **is right adjoint to** F, and write this $F \dashv G$) if for any objects X in $\mathcal{X}$ and A in $\mathcal{A}$ there is a given bijection $\varphi(X, A)$ between arrows from FX to A and arrows from X to GA,

$$\frac{FX \longrightarrow A}{X \longrightarrow GA}$$

which is moreover *natural in X and A*. This means that the following holds. Suppose that $X' \xrightarrow{x} X$ in $\mathcal{X}$ and $FX \xrightarrow{f} A$, then

$$\varphi(X, A)(f) \circ x = \varphi(X', A)(f \circ Fx)$$

and the similar condition holds for arrows $A \xrightarrow{a} A'$:

$$\varphi(X, A')(a \circ f) = a \circ \varphi(X, A)(f)$$

There are many examples of this concept in the text and exercises. For example, consider the the diagonal functor $\Delta : \mathcal{S} \longrightarrow \mathcal{S} \times \mathcal{S}$, which sends a set X to the pair (X, X) (and a mapping f to (f, f)). The *sum* functor $+ : \mathcal{S} \times \mathcal{S} \longrightarrow \mathcal{S}$, which sends (A, B) to $A + B$ is left adjoint to Δ, whereas the *product* functor is right adjoint to Δ. Another key example is the exponential or mapping-space functor $(-)^B$ whose properties follow from its definition as the right adjoint of $(- \times B)$.

In case F_1 and F_2 are both left adjoint to G, there is a natural isomorphism $F_1 \cong F_2$, and, similarly, any two right adjoints to F are naturally isomorphic.

Algebraic Topology:

The cohesive spaces of various categories (continuous, smooth, analytic, combinatorial) serve as domains for cohesively variable quantities and as arenas for uninterrupted motion. But in particular, a space also has qualitative attributes that collectively might be called its "shape," the best-known such attribute being measured by the Euler characteristic and the Betti numbers that count the number of k-dimensional "holes" in the space. (The qualitative attributes of a cohesive space can make a crucial difference when dealing with the differential and integral calculus of quantities that vary over such a space as domain; such quantities arise in electromagnetism and statistical mechanics, and the precise effect of the shape on their behavior is expressed by theorems of de Rham and of Stokes.) Not only Betti and Volterra, but also Vietoris and Noether, as well as Hurewicz and Steenrod, and many others, devised algebraic and combinatorial constructions for getting at these qualitative attributes. Those constructions were varied and sometimes complex so questions naturally arose: Do they always give the "same answer," and can they be adequately conceived in a more direct manner? In 1952 Eilenberg and Steenrod showed how to answer those questions by unifying the subject in a way that also made possible the ensuing 50 years of remarkable advance. Their use of the axiomatic method resulted from participating in the on-going development, extracting the essential features, and making these features explicit in order that they could serve as a basis and guide for further work in the field.

The analogy between set theory and algebraic topology goes even further. Namely, in both subjects the examples first considered satisfy a special axiom concerning the one-point space 1: In the abstract-set categories there are enough maps from 1 to separate maps between any two sets, and in singular cohomology, the cohomology of 1 vanishes in higher dimensions. But later important uses of both axiom systems have involved mathematically rich situations in which those special axioms do not hold. When it was first realized that these special axioms were too restrictive, one spoke of "generalized elements", "generalized cohomology theories," respectively but as the central role of these situations became established, the word "generalized" was gradually dropped, and now we speak just of "elements" (i.e., figures of various basic shapes, not just punctiform) and of "cohomology theories" (which do not necessarily vanish in higher dimensions even for the one-point space). In the case of set theory we approach these more general situations through our investigations of variable sets.

Category Theory:

Category Theory was made explicit by Eilenberg and Mac Lane. In 1945 [EM45] they concentrated some of the essential general features of the developments up to then in algebraic topology and in functional analysis, which are still-developing branches of the science that finds and uses the relation between quality and quantity in the study of space and number. A further development of category theory came in 1958 [Kan58] when Kan made explicit the notion of adjoint functors, which were then rapidly seen to be ubiquitous (even if implicit) in mathematics. In the 1960s and 1970s it was established in detail that the mathematically useful portions of set theory and of logic in the narrow sense can be seen as part of category theory [La69b], [KR77], [FS79].

Characteristic Functions (2.20):

Suppose we have fixed an element $1 \xrightarrow{v_1} V$ of a set V. (In the case of abstract sets we will often take V to be a two-element set and call the distinguished element "true".) Then a part i of a set A is said to have a map $A \xrightarrow{\varphi} V$ as **characteristic function** iff the elements of A, which are members of i, are precisely those to which φ gives the value true. In symbols,

$$\forall T \xrightarrow{a} A \; [a \in i \iff \varphi a = v_1 T]$$

(here $v_1 T$ denotes the composite map $T \longrightarrow 1 \xrightarrow{v_1} V$, which is constantly true). That is, $i = \varphi^{-1}[v_1]$ is the inverse image of the one-element part "true" of V along φ. In the category of sets, *every* part i has a *unique* characteristic function φ (to $V = 2$), and for any map f the two ways of substituting f agree:

$$\varphi_{f^{-1}[j]} = \varphi_j f$$

Composition:

Value of the Composite Mapping / Associative Law

$$\forall x [(gf)x = g(fx)]$$

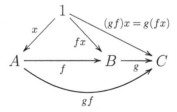

$T = 1$ elements/values

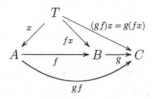

$$T = \text{arbitrary set : the associative law}$$

Contrapositive:

The **contrapositive of a statement** "*P* implies *Q*" is the statement that

"not *Q* implies not *P*" or in symbols: $(\neg Q \Rightarrow \neg P)$

   If the original implication is true, then the contrapositive is always true also, and in classical logic if the contrapositive is true, we can conclude that the original statement was true (the latter procedure being often referred to as proof by contradiction). For example: The statement that "1 is a separator in the category of sets" in its contrapositive guise assumes the following form (which is sometimes understood as the definition of equality for mappings):

$$[\forall\ 1 \xrightarrow{x} X\ [f_1 x = f_2 x]] \Rightarrow f_1 = f_2$$

Of course, to recognize the expression above as the contrapositive of the separator condition involves one more ingredient beyond the notion of contrapositive itself, namely, the recognition that existentially quantified statements are negated as follows:

$$\neg \exists x\, P \equiv \forall x \neg P$$

Converse of an Implication:

The **converse of a statement** of the form

P implies *Q* (or in symbols: $P \Rightarrow Q$)

is the statement

Q implies *P* (or in symbols: $Q \Rightarrow P$)

Even when the original statement is true, the converse statement may or may not be true, depending on the case.

Element (1.3):

- (in the narrow sense) An **element** of a set A is any mapping whose domain is 1 and whose codomain is A. This mapping is also sometimes referred to as the indication of an element, but we will not distinguish between elements and their indication.
- (in the generalized sense) An element of A is just another word for any mapping with codomain A; such a **generalized element** may also be referred to as a **variable element** or as a **parameterization** or **listing** of elements.

Epimorphism / Right Cancellation (4.5):

This concept is strictly "dual" to that of monomorphism. We say f is an **epimorphism** iff it satisfies the right-cancellation property

$$\forall \varphi_1, \varphi_2 \ [\varphi_1 f = \varphi_2 f \Rightarrow \varphi_1 = \varphi_2]$$

Foundation:

A foundation makes explicit the essential general features, ingredients, and operations of a science as well as its origins and general laws of development. The purpose of making these explicit is to provide a guide to the learning, use, and further development of the science. A "pure" foundation that forgets this purpose and pursues a speculative "foundation" for its own sake is clearly a nonfoundation.

Foundation, Category of Categories as:

Among the general features of mathematics a foundation should make explicit are

(1) the nature and workings of the logical and algebraic *theories* within which we reason and calculate;
(2) the nature and workings of the *set-universes*, whether abstract and constant, or cohesive and variable, that we imagine as an objective or geometrical background;
(3) the nature and workings of the *interpretations*, known as structures or models, of theories into set-universes.

These requirements have forced mathematics toward the foundational view we have tried to convey in this glossary and in this book; this can be seen in the following way: The "category" of models of a given theory in a given set-universe must form something like a mathematical category since these structures are different realizations of the one single system of features required by the theory, and hence there must be ways of comparing different ones by "morphisms" or maps that "preserve" those features. Something of that sort applies roughly to concepts generally, but in mathematics it is quite definite. For example, consider a theory A of addition and multiplication, which in an appropriate background universe U has many models such as

$Q(1) =$ a system of constant quantities;
$Q(S) =$ a system of quantities varying over a region S of space; and
$Q(T) =$ a system of quantities varying over an interval T of time.

Then a specified motion $m : T \longrightarrow S$ will yield a morphism that we name $m^* :$ $Q(S) \longrightarrow Q(T)$, and a specified instant $t : 1 \longrightarrow T$ of time will yield a morphism $t^* : Q(T) \longrightarrow Q(1)$, such morphisms being structure-preserving maps for the theory A. The background universe U must also be a category since morphisms of spaces (or sets) are the ingredients in the morphisms like t^*, m^* of structures. A key discovery around 1962 (foreshadowed by the algebraic logic of the 1950s) was that a theory A is also a category! That is because the most basic operation of calculation and reasoning (which as it turned out uniquely determines the other operations) is substitution, and substitution correctly objectified is composition. (The symbolic schemes often called "theories" are a necessary but not uniquely determined apparatus for presenting theories [in the usual algebraic sense of presentation].) It seems that two things (like A and U) cannot be concretely related unless they can be construed as objects of the same category; fortunately, in this case we have that both A and U are categories, and so a relation like $Q : A \longrightarrow U$ is a functor in a category of categories. Moreover, the morphisms between different models are just natural transformations between functors. These considerations put a mathematical focus on some questions that a foundation needs to address.

Function:

Although the word "function" is sometimes used in a way synonymous with "mapping," more frequently it is used to describe maps (in some category) whose codomains are some special objects V deemed to have a "quantitative" character (for example $V = 2$ or $V = \Omega$), and so one speaks of, for example, "the algebra of smooth complex-valued functions" on any smooth space. The V-valued functions on any object will always form an "algebra" (in a general sense) because the maps $V \longrightarrow V$, $V \times V \longrightarrow V$, and so on, will act as operations on them.

Functor (**10.18**):

Let $\mathcal{A}$ and $\mathcal{B}$ be categories. A **(covariant) functor** F, denoted $F : \mathcal{A} \longrightarrow \mathcal{B}$ from $\mathcal{A}$ to $\mathcal{B}$ is an assignment of

- an object $F(A)$ in $\mathcal{B}$ for every object A in $\mathcal{A}$
- an arrow $F(f) : F(A) \longrightarrow F(A')$ in $\mathcal{B}$ for every arrow $A \longrightarrow A'$ in $\mathcal{A}$

subject to the following equations:

- $F(1_A) = 1_{F(A)}$
- $F(gf) = F(g)F(f)$ whenever $A \xrightarrow{f} A' \xrightarrow{g} A''$

A **contravariant functor** $F : A \longrightarrow B$ assigns objects to objects, and an arrow $F(f) : F(A') \longrightarrow F(A)$ (note direction!) in B for every arrow $A \longrightarrow A'$ in A. It satisfies the equation $F(gf) = F(f)F(g)$ instead of the second equation above.

Inclusion (2.13):

The notation

$$i' \subseteq_A i$$

means *both* i, i' are parts of a set A and

$$\exists k \ [i' = ik]$$

which is equivalent to saying

$$\forall a[a \in i' \Rightarrow a \in i]$$

When the set A is understood, we write simply $i' \subseteq i$.

Intersection (2.36):

If i_1, i_2 are parts of A, then their **intersection** denoted

$$i_1 \cap i_2$$

is also a part j of A characterized up to equivalence by having as members precisely those elements of A *common* to i_1 and i_2

$$(\forall \ T \xrightarrow{a} A[a \in j \iff a \in i_1 \ \& \ a \in i_2]) \iff j \equiv i_1 \cap i_2$$

Among the many ways that intersection can be characterized is as the "inverse image" along i_1 of i_2.

Inverse Image (2.5):

If $A \xrightarrow{f} B$ is any mapping and j is any part of B, the condition "$f \in j$" is typically *not* outright true but does have a "solution set": the part of A for which $f \in j$ holds, more exactly the part i of A whose members are precisely all those elements a of

A for which $fa \in j$. In symbols:

$$\forall T \xrightarrow{a} A \; [a \in i \Longleftrightarrow fa \in j]$$
$$\Longleftrightarrow i \equiv f^{-1}[j]$$

We call $f^{-1}[j]$ the **inverse image** of j along f.

Injective / Monomap / Left-Cancellation / Part:

All four of these terms mean essentially the same thing but with different grammatical shadings as in "A *huge* person is a *giant*."

Injective (1.5):

Given an object I, we may say that f is I-**injective** iff f has the left-cancellation property for all test pairs with domain I:

$$\forall a_1, a_2 \; [I \underset{a_2}{\overset{a_1}{\rightrightarrows}} A \; \& \; fa_1 = fa_2 \Rightarrow a_1 = a_2]$$

Since the definition of "monomorphism" contains the phrase "$\forall T$," it is obvious that any monomorphism is always I-injective. The converse proposition, namely that I-injective $\Rightarrow$ monomorphism, holds if I is a **separator**, as is easy to prove. In the case of the category of abstract sets, we usually take $I = 1$ and refer to 1-injective mappings simply as injective. Thus **injective mappings** f are usually considered "by definition" to satisfy

$$fa_1 = fa_2 \Rightarrow a_1 = a_2$$

where a_1, a_2 are *elements* (in the narrow sense) of the domain of f; however, since 1 is a separator, injective mappings in fact satisfy the full left-cancellation property. In contrapositive form,

$$\forall a_1, a_2 [a_1 \neq a_2 \Rightarrow fa_1 \neq fa_2]$$

describes injective/monic f, thus to show that a certain f is **not** injective, it suffices to exhibit that

$$\exists a_1, a_2 \; [fa_1 = fa_2 \; \& \; a_1 \neq a_2]$$

Left-Cancellation / Monomapping (2.6):

The statement f **has the left-cancellation property** means that for any a_1 and a_2 for which $fa_1 = fa_2$ we can conclude $a_1 = a_2$, or in symbols

$$\forall T \; \forall a_1 \; \forall a_2 \; [fa_1 = fa_2 \Rightarrow a_1 = a_2]$$

as in the diagram

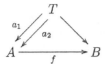

If f has the left-cancellation property in a certain category, we say that f is a **monomorphism** in that category. Since the morphisms in the category of sets are called mappings, the monomorphisms in that category may also be called **monomappings**.

Logic:

The science of logic in the ancient philosophical sense means the study of the general laws of the development of thinking. Thinking (1) reflects reality (i.e., has a content) but also (2) is itself part of reality and so has some motions that are oblivious to content. Therefore the science of logic finds two aspects of thought's motion: (1) the struggle to form a conceptual image of reality that is ever more refined, whose laws we may call objective logic, and (2) the motion of thought in itself (for example the inference of statements from statements), whose laws we may call subjective logic. Although grammar and some aspects of algebra might be considered as subjective logic, we will limit ourselves to the part we will sometimes refer to as logic in the narrow sense – that which is related to the inference of statements from statements by means dependent on their form rather than on their content. (Logic in the narrow sense is explained in more detail in Appendix A.)

Logic in the narrow sense is useful (at least in mathematics) if it is made explicit, and the work of Boole and Grassmann in the 1840s, Schröder in the late 1800s, Skolem in the early 1900s, Heyting in the 1930s (and of many others) has led to a high development, most aspects of which were revealed to be special cases of adjoint functors by 1970 [La69b]. The use of adjoint functors assists in reincorporating the subjective into its rightful place as a part of the objective so that it can organically reflect the objective and in general facilitate the mutual transformation of these two aspects of logic.

Logic, Objective:

The long chains of correct reasonings and calculations of which subjective logic is justly proud are only possible within a precisely defined universe of discourse, as has long been recognized. Since there are many such universes of discourse, thinking necessarily involves many transformations between universes of discourse as well as transformations of one universe of discourse into another. The results of applying logic in the narrow sense to the laws of these objective transformations are necessarily inadequate; for example, such attempts have led to the use of phrases such as "let X be a set in which there exists a group structure," which are essentially meaningless. Rather than using "there exists" in such contexts, one needs instead a logic of "given." Before category theory, at least one systematic discussion of the laws of these objective transformations was given by Bourbaki, who discussed how one structure could be deduced from another. The concepts of categories, functors, homomorphisms, adjoint functors, and so on, provide a rich beginning to the project of making objective logic explicit, but there is probably much more to be discovered.

Logic, Intuitionistic:

A. Heyting, in the 1930s, developed a logical algebra that happens to be applicable in any topos. This logic is weaker than the Boolean one because it allows for the possibility of motion or internal variation in the sets or universes to which it is being applied, and thus the typical Boolean inference (or subobject inclusion) "not(not A) entails A" is usually not correct. Because Heyting's work was intended to model the "constructivist" philosophy of the intuitionist Brouwer, Heyting's logic is sometimes incorrectly referred to as intuitionistic logic, and for that reason topos theory is sometimes referred to as constructive set theory; neither of these does justice to the true generality of Heyting's discovery or to Grothendieck's theory. There are a few toposes that have been specifically constructed in an attempt to understand the constructivist philosophy, but most toposes have little to do with it.

Mapping (1.1):

In the category of abstract sets, f is a **mapping** with domain A and codomain B if for *every* element x of A *there is exactly one* element y of B such that y is the value of f at x. In symbols,

$$\forall \ 1 \xrightarrow{x} A \ \exists!1 \xrightarrow{y} B[y = fx]$$

and as a diagram

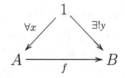

Membership (2.15):

The notation $a \in_A i$ means i is a part of a set A and $\exists \bar{a} \ [a = i\bar{a}]$ as in the diagram

When, as is usually the case, the set A is understood, we write simply $a \in i$.

Natural Transformation (10.19):

Let $F, G : \mathcal{A} \longrightarrow \mathcal{B}$ be functors. A natural transformation τ from F to G is the assignment of an arrow $\tau_A : FA \longrightarrow GA$ in $\mathcal{B}$ for each object A of $\mathcal{A}$ subject to the requirement that for each arrow $A_1 \xrightarrow{\alpha} A_2$ in $\mathcal{A}$ the following square commutes:

$$
\begin{array}{ccc}
FA_1 & \xrightarrow{\ \tau_{A_1}\ } & GA_1 \\
{\scriptstyle F\alpha}\big\downarrow & & \big\downarrow{\scriptstyle G\alpha} \\
FA_2 & \xrightarrow[\ \tau_{A_2}\ ]{} & GA_2
\end{array}
$$

The arrows τ_A are called the **components** of the natural transformation. In case each τ_A is an isomorphism in $\mathcal{B}$, then τ is called a **natural isomorphism**; in that case there is an inverse natural transformation.

There are many examples in the text and exercises – notably the natural map treated in Section 8.3, the singleton (see Exercise 8.9), and the arrows in the categories of $\mathcal{A}$-sets in Section 10.2.

One-Element Set (Section 1.2):

The set 1 is characterized by the fact that for any set A there is exactly one mapping with domain A and codomain 1. In symbols,

$$\forall A \ \exists! \ [A \longrightarrow 1]$$

that is, the statements below are both true:

$$\forall A \left[\begin{array}{l} \text{EXISTENCE } (\exists) \quad \exists f \ [A \xrightarrow{f} 1] \\[2mm] \text{AT MOST ONE } (!) \ \forall f, g \ [A \underset{g}{\overset{f}{\rightrightarrows}} 1 \Rightarrow f = g] \end{array} \right.$$

Part (2.11):

By a **part of a set** A is meant any monomapping i whose codomain is A. Notation: $U \xhookrightarrow{i} A$. Parts of A are called **equivalent** if they are isomorphic as objects of the slice category $\mathcal{S}/A$.

Separator (1.14):

A set T is a **separator** if there are enough mappings with it as domain to separate pairs of mappings with arbitrary domains; i.e. when A is an arbitrary set and if f_1 and f_2 are any two mappings with domain A and some common codomain B, if f_1 is distinct from f_2 there should exist a mapping x with domain T and codomain A such that $f_1 x$ is distinct from $f_2 x$. An important property of the category of abstract sets is that the terminal set 1 is a separator. In symbols,

$$\forall A \ \forall B \ \forall f_1, f_2 [[A \underset{f_2}{\overset{f_1}{\rightrightarrows}} B \ \& \ f_1 \neq f_2] \Rightarrow \exists \ 1 \xrightarrow{x} A[f_1 x \neq f_2 x]]$$

as in the diagram:

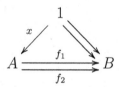

Set Theory:

Set theory was started in the late 1800s when Cantor made explicit an abstraction process (arising from his own work on Fourier analysis) and applied the idea of isomorphism (which he had learned from the work of Steiner on algebraic geometry) to the results of this abstraction process. Some of Cantor's followers did not fully appreciate this abstraction process (in editing Cantor's works for publication they issued judgments of it such as "kein glücklicher Versuch" and "um keinen Schritt weitergekommen," p. 35, edition of 1932 [C66]). For this and other reasons, too much of the technical development of set theory has been rather detached from its origins in the specific requirements of functional analysis and algebraic geometry. Thus, the great functional analyst and algebraic geometer of the 1950s and 1960s, Grothendieck, made only peripheral reference to set theory as known at the time when he devised topos theory, an explicit theory of the categories of cohesive and variable sets as they actually occur in geometry and analysis [AGV72]. In 1970 the essential part of Grothendieck's theory was drastically simplified in response to the needs of continuum mechanics and algebraic topology [La71], [T72] (see [Joh77],

[BW85], [MM92]). These developments made the bridging of the previous gap between axiomatic set theory and naive set theory possible.

Set Theory, Axiomatic:

The use of the axiomatic method, to make explicit the basic general features of the application of set theory to the study of mathematical subjects was delayed for much of the twentieth century by the attempt to popularize a certain philosophical view; according to that view the intersection of any two unrelated sets should have a well-defined meaning. That contrasts with the phenomenon noted in practice that inclusions, and more generally, relations between sets, are effected by specific mappings. The description of this "cumulative hierarchy" came to be considered by some as the only possible sort of formalized set theory.

Set Theory, Naive:

A certain body of set-theoretical methods and results is required for the learning and development of analysis, geometry, and algebra. Because of the insufficiently analyzed belief that this body could not be made logically precise without the cumulative hierarchy and global inclusion, it became customary to describe it in a "nonformalized" manner known as "naive set theory" [H60]. One of the purposes of this book is to overcome this division between naive and axiomatic aspects by giving formal axioms sufficient to describe the applicable aspects of the set theory that has been developed by Dedekind, Hausdorff and their many successors. For an earlier version, see [La64].

Set Theory, Parameterization:

Many of the applications of naive set theory could be described as based on "parameterization": a set is used to parameterize some things, some mappings on the set are deemed to express relations among the things, calculation and reasoning are applied to these maps, and the result is used to guide our dealing with the original things. Since in particular set theory should be partly applicable to itself, several of the axioms can be motivated by the observation that they simply assert that there are enough sets to parameterize some basic set-theoretic "things," as in the following three cases.

A set T can parameterize

(1) elements of a set A, by means of a map $T \longrightarrow A$; a bijective parameterization can be achieved by taking $T = A$ and 1_A as the map;

(2) pairs of elements of A, B by means of a pair of maps $T \longrightarrow A$, $T \longrightarrow B$; here a bijective parameterization has $T = A \times B$ and the projections as the maps;

(3) maps $A \longrightarrow B$ by means of a map $T \times A \longrightarrow B$; a bijective parameterization is possible if we can take $T = B^A$ and evaluation as the map.

Such parameterizations are the subject of the related ideas of universal mapping property, representable functors, and adjoint functors developed by Grothendieck, Kan, and Yoneda in the late 1950s.

(Sets [or spaces] themselves are T-parameterized in geometry by using a map $E \longrightarrow T$ with the fibers E_t being the sets parameterized; Cantor's diagonal argument showed that no single T can parameterize all sets.)

Surjective (1.4):

This concept is *not* logically dual to injective since it is an existential condition rather than one of cancellation. Hence, the *theorem* that for some categories "surjective" and "epimorphic" coincide will tell us something special about those categories. The exact notion of "surjective" is related to a chosen object I (such as $I = 1$ in the case of abstract sets). We say f is I-**surjective** iff

$$\forall y \exists x [fx = y]$$

where x, y are both supposed to have domain I but respective codomains A, B. That is, every I-element of the codomain is required to be a value of f (for at least one x).

To show that epimorphism $\Rightarrow$ 1-surjective in the case of abstract sets, suppose that f has the right-cancellation property but that (aiming toward a proof by contradiction) there is a y that does not appear as a value of f. But then if we take as φ_1 the characteristic function of y and as φ_2 the constantly "false": $B \longrightarrow 2$, it follows that $\varphi_1 f = \varphi_2 f$ but $\varphi_1 \neq \varphi_2$, contradicting that f has right cancellation. Hence, there is no such y, that is, f is surjective.

Topos:

Since 1970 [La71], the geometrical constructions of Grothendieck in SGA 4 [AGV72] were unified with the constructions of the set theorists Cohen, Scott and Solovay [B77] by the observation that both are largely concerned with categories $\mathcal{X}$ that

(1) have finite products and in which, moreover, there exist
(2) for each object A, a right adjoint $(\ )^A$ to the functor $A \times (\ )$, and

(3) a map $1 \xrightarrow{\text{true}} \Omega$, which is a universal part in the sense that for any part $\alpha : A \hookrightarrow X$ of any object in $\mathcal{X}$, there is a unique $1 \xrightarrow{\varphi} \Omega$ such that

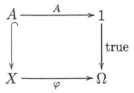

is a pullback; i.e. for any $T \xrightarrow{x} X$

$$x \text{ belongs to } \alpha \text{ iff } \varphi x = \text{true } T$$

This is sometimes called the "elementary" theory of toposes since the three axioms are finitary and internal to $\mathcal{X}$.

Topos and the Cantorian Contrast:

The cohesive and variable sets ("spaces") are persistent ideas collectively developed as part of our coping with the world of matter in motion of which we are an interacting part. Cantor and subsequent mathematicians have found it useful to apply something like the ancient Greek ideas of arithmos and chaos by analyzing this variation and cohesion through contrasting it with constancy and noncohesion $\mathcal{Y}$, wherein objects are distinct and fixed enough to have a property like number but nearly devoid of any other property. (Cantor further conjectured, in effect, that if "nearly devoid" could be idealized to "totally devoid" for a topos $\mathcal{Y}$ of abstract sets, then $\mathcal{Y}$ would satisfy some strong special properties such as the continuum hypothesis – a conjecture apparently proved in the 1930s by Gödel, although that interpretation is still not widely accepted.) Mathematical practice made clear that cohesion and variation occur in many diverse but related categories $\mathcal{X}$. Construing some of those categories as toposes makes it possible to study the contrast arising in Cantor's abstraction process as a geometric morphism $\mathcal{X} \xrightarrow{\Gamma} \mathcal{Y}$. This process Γ_* of extraction of pure points unites two opposite inclusions of $\mathcal{Y}$ into $\mathcal{X}$: the subcategory Γ^* of discrete spaces with zero cohesion and zero motion and the subcategory $\Gamma^!$ of codiscrete spaces with total (but banal) cohesion and motion. These opposite but identical subcategories provide a zeroeth approximation to the reconstruction of any space X by placing it canonically in an interval

$$\Gamma^* \Gamma_* X \longrightarrow X \longrightarrow \Gamma^! \Gamma_* X$$

(In fact closer approximations to X involving narrower intervals are often obtainable by geometric morphisms involving a less abstract topos $\mathcal{Y}'$, but we do not discuss that here.) There are no nonconstant $\mathcal{X}$-maps from a codiscrete space to a discrete space;

this sharply separates the two aspects (of a $\mathcal{Y}$-object) that the elements are totally distinguished and yet have total lack of distinguishing properties; these aspects are thus revealed to form not a conceptual inconsistency but a productive dialectical contradiction.

Topos, Geometric Morphism:

A pair of adjoint functors

$$\mathcal{X} \underset{\Gamma_*}{\overset{\Gamma^*}{\rightleftarrows}} \mathcal{Y}$$

with Γ^* left adjoint to Γ_* is called a geometric morphism $\mathcal{X} \xrightarrow{\Gamma} \mathcal{Y}$ in case Γ^* preserves finite limits. The latter "exactness" property will automatically be true in case there is a further left adjoint $\Gamma_!$ to Γ^*. (Note that this string $\Gamma_! \dashv \Gamma^* \dashv \Gamma_*$ is in objective analogy with the string $\exists \dashv$ substitution $\dashv \forall$ of logic in the narrow sense.) When $\Gamma_!$ exists, the morphism Γ is often called "essential". Grothendieck's compact notation is: asterisks for functors that are present for every Γ, exclamation points for functors that are present only for notably special Γ, whereas the lower position (for either) denotes functors deemed to have the same direction as Γ and the upper position denotes functors having the opposite direction to Γ. In case a further right adjoint $\Gamma^!$ to Γ_* exists, a special geometric morphism Φ results in the opposite direction with $\Phi^* = \Gamma_*$ and $\Phi_* = \Gamma^!$; such a Φ may be called "flat" because the left adjoint $\Phi_!(= \Gamma^*)$ not only exists but preserves finite limits.

All four functors exist in the case $\mathcal{X} = \mathcal{Y}^{\Delta_1^{op}}$, the category of reflexive graphs in a topos $\mathcal{Y}$ (see Section 10.3), where $\mathcal{X} \xrightarrow{\Gamma} \mathcal{Y}$ is determined by $\Gamma^* = $ the inclusion of trivial graphs.

Topos of Abstract Sets:

For a topos $\mathcal{U}$ the conditions that
(1) epimorphisms have sections (axiom of choice), and
(2) there are (up to isomorphism) exactly two parts of 1,
are generally sufficient to exclude most traces of cohesion or variation, as Cantor proposed. (It follows that $2 = \Omega$ and that 1 separates.)

Occasionally the (provably stronger) Generalized Continuum Hypothesis of Cantor may be further imposed:

(GCH) Whenever monomorphisms $A \longrightarrow X \longrightarrow 2^A$ exist, then an isomorphism
$$A \cong X \text{ or } X \cong 2^A \text{ also exists.}$$

Usually a topos of abstract sets is assumed to satisfy (1) and (2) and also Dedekind's axiom concerning the existence of a set of natural numbers:

(3) The forgetful functor $\mathcal{U}' \longrightarrow \mathcal{U}$ has a left adjoint, where $\mathcal{U}'$ is the category whose objects are the objects of $\mathcal{U}$ equipped with an action by an arbitrary endomap.

Topos, Grothendieck:

For the development of number theory, algebraic geometry, complex analysis, and of the cohomology theories relevant thereto, it was found necessary in 1963 to introduce explicitly certain categories that Grothendieck and Giraud called "$\mathcal{U}$-toposes." ("Topos" is a Greek term intended to describe the objects studied by "analysis situs," the Latin term previously established by Poincaré to signify the science of place [or situation]; in keeping with those ideas, a $\mathcal{U}$-topos was shown to have presentations in various "sites".)

In modern terms a $\mathcal{U}$-topos is a topos $\mathcal{X}$ equipped with a geometric morphism Γ to a topos $\mathcal{U}$ of abstract sets satisfying the condition that $\mathcal{X}$ is boundedly generated over $\mathcal{U}$ in the sense that some single map $B \longrightarrow \Gamma^* I$ generates $\mathcal{X}$. In the cases where one can choose $I = 1$, the generation condition means that the map

$$\mathcal{X}(B, X) \cdot B \longrightarrow X$$

is an epimorphism for all X, where by definition

$$\mathcal{X}(B, X) = \Gamma_*(X^B)$$
$$U \cdot B = (\Gamma^* U) \times B$$

The functor $\Gamma_* = \mathcal{X}(1, -)$ is sometimes called the "global sections" functor to remind us that the objects of $\mathcal{X}$ may be spatially more variable than those of $\mathcal{U}$, or the "fixed points" functor to remind us that the objects of $\mathcal{X}$ may be dynamically more active, or the "underlying discrete set" functor to remind us that they may be more cohesive than those of $\mathcal{U}$.

As an application of Grothendieck's powerful method of relativization, it is very useful to consider $\mathcal{U}$-topos also in the more general case where $\mathcal{U}$ itself is an arbitrary topos not necessarily consisting of purely abstract sets.

Union:

The hierarchical philosophy led to the identification of the notions of set and of subset (part). This in turn led to a confusion between sums (of sets) and unions (of parts of a given set); for example, the editors of Cantor's collected works [C66] complained that he had defined the sum only in the case of disjoint sets. If $A_i \longrightarrow X$ is a given family of parts (for example a family consisting of only two parts), then the **union of the family** is a part of X obtained by (epi–mono) image factorization of the induced map from the sum

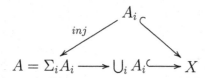

In toposes, any epimorphism is a coequalizer of the induced equivalence relation given by the self-pullback $A \times_X A \rightrightarrows A$. In turn this relation can be expressed as

$$\Sigma_{i,j} A_i \times_X A_j \rightrightarrows \Sigma A_i$$

(where each summand $A_i \times_X A_j$ is an intersection of parts and so is again a part of X). By the universal mapping property of coequalizers, any part $B \longrightarrow X$ to which each of the given parts belongs will also contain the union; i.e. in the category of parts of X

$$\frac{\bigcup A_i \subseteq B}{A_i \subseteq B, \text{ for all } i}$$

are equivalent; the objective form of the rules of inference for disjunction (= repeated "or"). However, the universal mapping property of the union is stronger than that since even if B is an arbitrary object (not necessarily the domain of a part of X), any given family $A_i \xrightarrow{\beta_i} B$ of maps (not necessarily monos) will be derivable from a unique single map $\bigcup_i A_i \longrightarrow B$ provided only that the equations

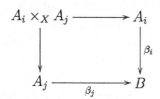

expressing agreement on the overlap are satisfied by the given family β_i.

Yoneda Embedding:

A functor $Q : \mathcal{A} \longrightarrow \mathcal{S}$ is called **representable** if there is an object A in $\mathcal{A}$ and an element q of the set $Q(A)$ so that for all B in $\mathcal{A}$, the map $\mathcal{A}(A, B) \longrightarrow Q(B)$, which takes any $A \xrightarrow{b} B$ into $Q(b)(q)$, is an isomorphism of sets. Equivalently, there exists a natural isomorphism of functors $\mathcal{A}(A, -) \cong Q$, where $\mathcal{A}(A, -)$ is the very special functor $A \xrightarrow{\mathcal{A}(A,-)} \mathcal{S}$ whose values "are" (i.e., perfectly parameterize) the maps with domain A provided by the category $\mathcal{A}$. It is a consequence of Yoneda's lemma that any two representing objects for the same functor Q are isomorphic, and indeed there is a unique isomorphism a between them such that the induced $Q(a)$ takes the "universal element" q given with one representation into the universal element given with the other. For any map a (not necessarily an isomorphism) from A_1 to A_2 in a category $\mathcal{A}$, the composition of maps induces a corresponding natural transformation $a^* : \mathcal{A}(A_2, -) \longrightarrow \mathcal{A}(A_1, -)$ (see Exercise 10.22) in the opposite direction between the corresponding representable functors. Thus, there is a functor

$$\mathcal{A}^{\mathrm{op}} \xrightarrow{Y} \mathcal{S}^{\mathcal{A}}$$

called the **Yoneda embedding**, defined by $Y(A) = \mathcal{A}(A, -)$.

The very same construction applied to the opposite category yields the notion of contravariant representable functor $\mathcal{A}^{\mathrm{op}} \xrightarrow{X} \mathcal{S}$ as one for which $X(B) \cong \mathcal{A}(B, A)$ naturally for all B but for a fixed A that is said to **represent** X. Because the resulting Yoneda embedding

$$\mathcal{A} \xrightarrow{Y} \mathcal{S}^{\mathcal{A}^{\mathrm{op}}}$$

is full and faithful (see Yoneda's lemma), one often suppresses any symbol for it and considers $\mathcal{A}$ to be simply a special subcategory of $\mathcal{S}^{\mathcal{A}^{\mathrm{op}}}$; and conversely, subcategories $\mathcal{C}$ of $\mathcal{S}^{\mathcal{A}^{\mathrm{op}}}$ that contain this $\mathcal{A}$ often embody very useful generalizations of the concept that $\mathcal{A}$ itself embodied. The special cases in which $\mathcal{A}$ is a monoid or a poset were already studied and used over a century ago: Cayley observed that a canonical example of a left action of $\mathcal{A}$ is the action of $\mathcal{A}$ itself on the right, and Schröder, Dedekind, Cantor, and others made much use of the canonical representation of a poset as inclusions between certain subsets (actually right ideals) of a given set (namely of its own set of objects). Note that the category **2** with one nonidentity arrow (see Section 6.2) embeds in $\mathcal{S}$ as the sets 0 and 1 and that the Yoneda embedding of a given category $\mathcal{A}$ actually belongs to the induced subcategory

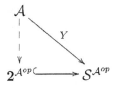

if and only if $\mathcal{A}$ is a poset.

Yoneda's Lemma:

For any contravariant functor $\mathcal{A}^{\mathrm{op}} \xrightarrow{X} \mathcal{S}$ to sets and for any given object A, the value $X(A)$ is isomorphic as a set to the (a priori "big") set of natural transformations $\mathcal{A}(A, -) \longrightarrow X$ (see Exercise 10.24). When we introduce some still larger categories via exponentiation in the category of categories, wherein S is an object, this says more fully that the functor "evaluate at A"

$$S^{A^{op}} \longrightarrow S$$

is one of the representable functors on $S^{A^{op}}$, indeed it is representable *by* the representable functor on A^{op}. Corollaries of Yoneda's lemma include the fullness of the Yoneda embedding, the uniqueness of representing objects, and the fact that "any" set-valued functor is canonically a colimit of representable ones. In case A itself is **small** relative to S (which means that the objects of A can be parameterized by an object of S and that A has small hom-sets), then the latter is equivalent to the fact that any X has a presentation as a coequalizer of two maps between two infinite sums of representable functors.

Yoneda and Totality:

For any category A such that the arrows between any two objects of A can be parameterized by an object of S, we say that A has **small hom-sets**.

A category A with small hom-sets is **totally cocomplete** if its Yoneda embedding $Y_* = Y$ has a left adjoint $Y^* : S^{A^{op}} \longrightarrow A$. For example, the category of groups or the category $A = S^{C^{op}}$ of C-sets are totally cocomplete; Y^* too may have a left adjoint $Y_!$. For example, C-sets, but not groups, have such an "opposite" embedding $Y_!$ into the same functor category. One and only one category A has both a further adjoint to $Y_!$ and also a still further left adjoint to that, i.e., an adjoint string of length 5

$$U \dashv V \dashv Y_! \dashv Y^* \dashv Y_*$$

starting with $Y_* = Y$ on the right, relating A to $S^{A^{op}}$, namely $A \cong S$, the base category of sets itself [RW94]. There are several subcategories of the functor category $S^{S^{op}}$ that still unite the opposites $Y_!$ and Y_*; for example, the category of simplicial complexes that are important in combinatorial topology. Another subcategory of $S^{S^{op}}$ containing Y_* as the codiscrete objects is the category of bornological sets, important in functional analysis, where $X : S^{op} \longrightarrow S$ is considered as a structure in which $X(S) \cong \mathrm{Bor}(S, X)$ acts as the set of "bounded" S-parameterized parts of X.

C.2 Mathematical Notations and Logical Symbols

$\longrightarrow$ denoting a map (also called transformation or function) from a **domain** A to a **codomain** B, $A \longrightarrow B$

$\Rightarrow$ $A \Rightarrow B$ (A implies B) or (A only if B) or (if A then B)

$\Leftarrow$ $A \Leftarrow B$ (A is implied by B) or (A if B) or (if B then A)

$\Leftrightarrow$ $A \Leftrightarrow B$ also written "iff" (A if and only if B)
$A \Rightarrow B$ & $B \Rightarrow A$ (A implies B and B implies A)

$\longmapsto$ value at

$\overset{\sim}{\longrightarrow}$ a map that has an inverse (called **isomorphism**)

$\hookrightarrow$ representing the domain as a **part** of the codomain

$=$	equals	$\neq$	not equal
$\equiv$	equivalence of parts	$\subseteq$	inclusion (of a part in another part)
$\forall$	for all		
$\exists$	there exists (existential statement)	$!$	unique (= exactly one)
		$\neg$	not
$\in$	membership (a member of)	$\cap$	intersection (of parts)
$>$	greater than	$<$	smaller than
∞	infinity	Σ	sum
$\vdash$	entails	$\int$	integral sign
$\wedge$	and, also &	$\vee$	or
fg	composite of two maps (read "f following g")	$\langle \ \rangle$	notation for ordered pairs or triples or n-tuples
$\backslash$	but not	∂	boundary of

C.3 The Greek Alphabet

Since the Latin letters a, b, x, y, and so forth, are sometimes not enough to express all the different mathematical entities that may arise in a given discussion, Greek letters are also used in all fields where mathematics is applied. Usually their meaning is specified at the beginning of each discussion just as is done for the meaning of the Latin letters. The Greek letters themselves are pronounced as follows:

α	alpha		ν	nu
β	beta		ξ	xi
γ	gamma		o	omicron
δ	delta		π	pi
ϵ	epsilon		ρ	rho
ζ	zeta		σ	sigma
η	eta		τ	tau
θ	theta		υ	upsilon
ι	iota		φ	phi
κ	kappa		χ	chi
λ	lambda		ψ	psi
μ	mu		ω	omega

This list contains the lowercase Greek letters. The uppercase sigma and pi are often used to denote sum and product and are written Σ and Π. Other frequently occurring uppercase Greek letters are Ω (omega), Δ (delta), Γ (gamma), Ψ (psi), Φ (phi), and $\mathcal{X}$ (chi).

Bibliography

[AGV72]   M. Artin, A. Grothendieck, and J. L. Verdier. *Théorie des topos et cohomologie étale des schémas, (SGA4)*. Volume 269 and 270 of *Lecture Notes in Math.*, Springer-Verlag, 1972.

[BW85]   M. Barr and C. Wells. *Toposes, Triples and Theories. Grundlehren Math. Wiss. 278*, Springer-Verlag, 1985.

[BW95]   M. Barr and C. Wells. *Category Theory for Computing Science, second edition*. Prentice-Hall, 1995.

[B77]   J. L. Bell. *Boolean-valued models and independence proofs in set theory*. Clarendon Press, 1977.

[B97]   J. L. Bell. Zorn's lemma and complete Boolean algebras in intuitionistic type theories. *Journal of Symbolic Logic*, 62:1265–1279, 1997.

[B99]   J. L. Bell. *A Primer of Infinitesimal Analysis*. Cambridge University Press, 1998.

[C66]   G. Cantor. Abhandlungen mathematischen und philosophischen Inhalts. Ed: Ernst Zermelo. Georg Olms, 1966. (Reprografischer Nachdruck der Ausgabe Berlin 1932.)

[D75]   R. Diaconescu. Axiom of choice and complementation. *Proceedings of the American Mathematical Society*, 51:176–178, 1975.

[EM45]   S. Eilenberg and S. Mac Lane. General theory of natural equivalences. *Transactions of the American Mathematical Society*, 58:231–294, 1945.

[ES52]   S. Eilenberg and N. Steenrod. *Foundations of Algebraic Topology*. Princeton University Press, 1952.

[FS79]   M. Fourman and D. S. Scott. Sheaves and logic. In *Proceedings of the Durham Symposium, 1977*. Lecture Notes in Mathematics No. 753, Springer-Verlag, 1979.

[G64]   H. Gaifman. Infinite boolean polynomials I. *Fund. Math.* 54: 229–
           250, 1964.

[H60]   P. Halmos. *Naive Set Theory.* Van Nostrand, 1960.

[Is60]   J. Isbell. Adequate subcategories. *Illinois J. Math* 4: 541–552, 1960.

[Joh77]   P. T. Johnstone. *Topos Theory.* Academic Press, 1977.

[JM95]   A. Joyal and I. Moerdijk. *Algebraic Set Theory.* Cambridge Univer-
           sity Press, *London Math Soc. Lecture Note Series 220,* 1995.

[Kan58]   D. Kan. Adjoint functors. *Transactions of the American Mathemati-
           cal Society,* 87:294–329, 1958.

[KR77]   A. Kock and G. Reyes. Doctrines in categorical logic. *Handbook of
           Mathematical Logic,* North-Holland, 1977.

[La64]   F. W. Lawvere. Elementary Theory of the Category of Sets. *Proceed-
           ings of the National Academy of Sciences* 52:1506–1511, 1964.

[La69a]   F. W. Lawvere. Diagonal arguments and cartesian closed categories.
           Lecture Notes in Math, 92:134–145, Springer-Verlag, 1969.

[La69b]   F. W. Lawvere. Adjointness in foundations. *Dialectica,* 23:281–296,
           1969.

[La71]   F. W. Lawvere. Quantifiers and sheaves. In *Actes du congrès inter-
           national des mathématiciens,* 329–334, Gauthier-Villars, 1971.

[La73]   F. W. Lawvere. Metric spaces, generalized logic and closed cate-
           gories. *Rendiconti del Seminario Matematico e Fisico di Milano,*
           43:135–166, 1973.

[LaS97]   F. W. Lawvere and S. H. Schanuel. *Conceptual Mathematics. A first
           introduction to categories.* Cambridge University Press, 1997.

[MM92]   S. Mac Lane and I. Moerdijk. *Sheaves in Geometry and Logic.*
           Springer-Verlag, 1992.

[M92]   C. McLarty. *Elementary Categories, Elementary Toposes.* Oxford
           University Press, 1992.

[M87]   E. Mendelson. *Introduction to Mathematical Logic.* Third edition.
           Wadsworth and Brooks/Cole, 1987.

[RW94]   R. Rosebrugh and R. J. Wood. An adjoint characterization of the
           category of sets. *Proceedings of the American Mathematical Society,*
           122:409–413, 1994.

[Sup72]   P. Suppes. *Axiomatic Set Theory.* Dover, 1972.

[T72]   M. Tierney. Sheaf theory and the continuum hypothesis. *Lecture
           Notes in Math,* 274:13–42, Springer-Verlag, 1972.

[Wal91]   R. F. C. Walters. *Categories and Computer Science.* Cambridge Uni-
           versity Press, 1991.

Additional References

G. Boole (1815–1864). *Collected Logical Works (2 volumes)*. Chicago: Open Court Publ. Co., 1916 and 1940.

G. Cantor (1845–1918). *Contributions to the Founding of the Theory of Transfinite Numbers*. La Salle, Ill.: Open Court Publ. Co., 1952.

R. Dedekind (1831–1916). *Gesammelte mathematische Werke*. Braunschweig, 1930.

G. Galilei (1564–1642). *Discorsi e dimostrazioni matematiche intorno a due nuove scienze attinenti alla meccanica e i movimenti locali*. Leiden: 1638. ("Discourses on Two New Sciences")

K. Gödel (1906–1978). *The Consistency of the Axiom of Choice and of the Generalized Continuum-Hypothesis with the Axioms of Set-Theory. Annals of Mathematics Studies* No. 2. Princeton, N.J.: Princeton University Press, 1940.

H. G. Grassmann (1809–1877). *Die Ausdehnungslehre, ein neuer Zweig der Mathematik*. Gesammelte Werke, Ersten Bandes erster Theil (Leipzig: B. G. Teubner 1894) (Reprint Bronx, N.Y.: Chelsea, 1969).

A. Grothendieck (born 1928). *Techniques des constructions en géométrie analytique. Séminaire Henri Cartan*, 11:1–28, 1960–61.

F. Hausdorff (1869–1942). *Das Chaos in kosmischer Auslese, ein erkenntniskritischer Versuch von Paul Mongré (pseud.)*. Leipzig: C. G. Naumann, 1898.

F. Hausdorff. *Grundzüge der Mengenlehre*. Leipzig: Von Veit, 1914.

A. Heyting (1898–1980). *Mathematische Grundlagenforschung: Intuitionismus, Beweistheorie*. Ergebnisse der Mathematik und ihrer Grenzgebiete 3–4. Berlin: Springer, 1934.

E. Schröder (1841–1902). *Vorlesungen über die Algebra der Logik (exakte Logik)*. Abriss der Algebra der Logik, Bronx, N.Y.: Chelsea Publ. Co., 1966.

T. Skolem (1887–1963). *Selected Works in Logic*. Edited by Jens Erik Fenstad. Scandinavian University Books, Oslo: Universitetsforlaget, 1970.

J. Steiner (1796–1863). *Gesammelte Werke*. Herausgegeben von K. Weierstrass auf Veranlassung der königlich-preussischen Akademie der Wissenschaften, Berlin: G. Reimer, 1881–82.

E. F. Zermelo (1871–1953). Untersuchungen über die Grundlagen der Mengenlehre: I. *Math. Annalen* 65:261–281, 1908.

M. Zorn (1906–1993). A remark on method in transfinite algebra. *Bull. Amer. Math. Soc.* 41:667–670, 1935.

Index

Definitions appear in **bold** type.

257

Printed in the United States
By Bookmasters